SOLID WASTE MANAGEMENT IN RURAL AREAS

Edited by **Florin-Constantin Mihai**

Solid Waste Management in Rural Areas
http://dx.doi.org/10.5772/66551
Edited by Florin-Constantin Mihai

Contributors

Qiuyan Yuan, Marco Nuti, María Cristina Echeverria, Elisa Pellegrino, Petra Schneider, Anh Hung Le, Jan Šembera, Rodolfo Daniel Silva-Martínez, Christia Meidiana, Harnenti Afni Yakin, Wawargita Permata Wijayanti, Cácio Boechat, Adriana Arauco, Rose Duda, Antonny Sena, Manoel Souza, Ana Brito, Iria Villar, David Alves, Salustiano Mato, Xosé Manuel Romero, Bernardo Varela, Sandro Xavier De Campos, Rosimara Zittel, Luciléia Granhen Tavares Colares, Karine Marcondes Da Cunha, Tatiana Windékpè KOURA, Valentin Missiakô KINDOMIHOU, Gustave Dieudonné DABENONBAKIN, Brice Augustin SINSIN, Florin-Constantin Mihai, Mohammad Taherzadeh

Published by InTech
Janeza Trdine 9, 51000 Rijeka, Croatia

Publishing Process Manager Mirena Calmic
Technical Editor SPi Global
Cover InTech Design team

First published September, 2017
Printed in Croatia
Legal deposit, Croatia: National and University Library in Zagreb

Additional hard copies can be obtained from orders@intechopen.com

Solid Waste Management in Rural Areas, Edited by Florin-Constantin Mihai
p. cm.
Print ISBN 978-953-51-3485-5
Online ISBN 978-953-51-3486-2

For EU product safety concerns:
IN TECH d.o.o., Prolaz Marije Krucifikse Kozulić 3, 51000 Rijeka, Croatia,
info@intechopen.com or visit our website at intechopen.com.

Meet the editor

Florin-Constantin Mihai holds a PhD degree in Geography from the Department of Geography, "Alexandru Ioan Cuza" University of Iasi (Romania), and BSc and MSc degrees in Environmental Science. He published papers on various topics regarding environmental and waste management issues. He promotes the geography of waste as a new complementary approach in the environmental and social sciences. His research aims to develop new methods and waste indicators in order to assess the key waste management issues across various geographical scales, particularly in transition and developing countries. He has a particular interest in rural waste management sector.

Contents

Foreword

The human population is growing continuously, and now we are more than 7 billion people in the world. In addition, the global welfare has been improved in the last decades, which results in producing more of our resources and producing more wastes in the world. Waste management in urban areas has been on special focus for several reasons. However, according to the World Bank, still, 46% of the population in the world is living in the rural areas. It is therefore important to make a special attention to the waste management in the rural areas, which is the topic of this book.

A proper waste management and resource recovery demands several factors to consider including like the technical, social, legal, environmental, and economical factor. These factors in the rural and urban area are quite different, and, therefore, a proper method of waste management in rural and urban areas varies from each other. As example, the financing of waste management in the cities do usually exist via direct or governmental taxes, while the economy of rural areas does not facilitate the full cost of treating wastes. On the other hand, the willingness and engagement of people in cleaning the villages and their relationships are different than the cities where personal responsibility in waste management is usually low, and organizations such as municipalities or private companies are responsible for treating the wastes. In addition, while the amounts of wastes in large cities are collected in special areas and it could be economically feasible to treat it properly, the distribution of solid wastes in rural areas is a hinder in this aspect.

This book refers to the waste management of rural areas. In addition to the general discussion on this topic, there are different chapters dedicated to social and technical aspects that are presented in a form of case studies. Waste sorting at source (household separation) and people's willingness and also the function of scavengers and unofficial system to separate recycling materials are discussed. In terms of technical aspects, the decentralized and home composting and anaerobic digestion can be mentioned and are discussed in separate chapters. These technologies can convert the wastes to value-added products that can be used in rural areas. Compost or digestates are examples of the composting or anaerobic digestion products that are usually proper for improving agricultural soils, where other products such as biogas can be used for cooking, heating, and/or gas lights in household scales and electricity production in larger scales. These aspects are discussed in different chapters.

In addition to the municipal wastes, agricultural wastes are usually present in rural areas. This is a specific aspect of rural areas. Therefore, the treatment of agricultural wastes separately or together with municipal wastes in rural areas is an important factor to consider.

Best regards,
Mohammad J. Taherzadeh
University of Borås, Sweden

Preface

Solid waste management in rural areas is a key issue in developing and transitioning countries due to the lack of proper waste management facilities and services. Even several high-income countries have serious challenges because of landfill-based approach in regional and local waste management systems.

The book points out the current gaps in rural waste management sector and the new opportunities in the management of municipal and agricultural waste.

The book contains nine chapters with worldwide contributions, which are discussed starting from municipal solid waste fraction toward agricultural sources.

The first chapter (introductory chapter) performs a critical overview of rural waste management sector at the global level highlighting, on the one hand, the public and environmental threats associated with improper waste disposal practices, and on the other hand, it outlines the two main routes toward waste prevention and rural sustainability.

The second chapter examines the community acceptance for household separate collection (dry recyclables) in order to increase the rural waste bank's effectivity, an emerging public community participation in dealing with solid waste management issues in Indonesia.

The third chapter investigates the waste management activities performed by informal sector across rural areas with social and environmental implications. The study has a global coverage where the role of informal sector activities from high-, middle-, and low-income countries is analyzed: Austria, the Czech Republic, Germany, Jordan, Mexico, Nepal, South Africa, and Vietnam.

The fourth chapter points out the role of decentralized composting activities in a rural region of Galicia (Spain) in the recovery process of biowaste fraction supporting a sustainable local development and, on the other hand, the waste diversion rate from landfills as promoted by EU waste hierarchy framework.

The fifth chapter reveals the positive and negative effects of using organic waste as soil conditioners or fertilizers in agriculture for increasing crop productivity. The chapter highlights the necessity of proper monitoring regarding the environmental risks and public health threats associated with the use of organic waste on agricultural lands.

The sixth chapter proposes an alternative for the home composting procedure using a facultative reactor. The experimental results show that such approach is a promising technology to manage organic solid residue at a rural household scale.

The seventh chapter supports the anaerobic digestion as one of the best available techniques for treatment of organic waste fraction of municipal waste based on a critical literature review. This biotechnology can be further improved, and it is suitable for rural communities.

The eighth chapter analyzes the linkage between improvement of palm oil process extraction and palm oil mill solid waste management for sustainable palm oil production. Current status and future prospects of this agricultural waste fraction are further examined in rural areas of Benin Republic.

The ninth chapter performs a depth analysis concerning the rural waste management perspectives in the case of two important agricultural sources such as coffee bean production (Latin America) and the extraction process of olive oil (Mediterranean countries) supported by a relevant literature review.

The book examines major topics and issues which rural communities are nowadays facing concerning municipal and agricultural wastes. The book will provide useful information for academics, various professionals, members of civil society, and national and local authorities.

The book points out that rural regions need proper attention at the global level concerning solid waste management sector.

Dr. Florin-Constantin Mihai
Department of Research
Faculty of Geography and Geology
"Alexandru Ioan Cuza" University
Iasi, Romania

Chapter 1

Introductory Chapter: Rural Waste Management Issues at Global Level

Florin-Constantin Mihai and
Mohammad J. Taherzadeh

Additional information is available at the end of the chapter

http://dx.doi.org/10.5772/intechopen.70268

1. Introduction

Different technical and social innovations may be required for solid waste management sector in large cities and rural areas as particular geographical regions [1]. Despite the fact that dumps represent the worst-case scenario in current waste management practices in terms of environmental protection and sustainability, they still occurred across the globe, particularly in peri-urban and rural regions. Developing countries are facing the transition from the dumps to the implementation of the first sanitary landfills. Former communist countries are facing serious challenges in the closure of "conventional landfills" which do not meet the criteria of the EU Landfill Directive 1999/31. Some of these sites must be upgraded in order to comply the current EU standards, and new integrated waste management system must replace the obsolete infrastructure. Sweden, Denmark, and Germany have developed their waste management toward "zero waste landfill," while other countries such as the USA, India, Brazil, and Qatar still use landfilling as the main option in their waste management [2].

Developed, transition, and emerging countries did not eradicate the wild dump issues. Despite the fact that these sites are smaller than formal urban landfills and scattered across peri-urban and rural regions, they are still a significant pollution source. Wild dumps must be mapped at municipal level across all regions in order to assess their environmental impact [3, 4]. Monitoring of illegal dumping activities is crucial either in high-income countries affecting public lands, roadsides, or water bodies [5–7].

The dump is historically the basic and most convenient option in the waste management treatment used by human settlements across the globe along with ocean and river dumping practices.

The lack of governmental policy and finance, difficulty in political issues and long-term planning in waste management, social behavior, and resistance to change in, for example, separation of wastes at source, regular waste collection services, poor waste management infrastructure, low quality of waste management services, lack of funds, poor environmental awareness, low market for recycled materials, all these factors contribute to the existence of open dumps nowadays [2].

The wild dumps are encountered in the peri-urban and rural areas due to the lack of waste and sanitation facilities. Frequently, such uncontrolled disposal sites are located in the proximity of households and water bodies. The dumps are a source of complex pollution (air, water, soil, and biodiversity) which threatens the public health. Mixed waste fractions (municipal, agricultural, construction and demolition, WEEE, bulk items), including hazardous streams, are disposed in such sites causing serious public health issues.

In some cases, such dumps are heavily pollution source due to the illegal disposal activities practiced by the mafia in southern Italy (so-called mob dumping). Particular geographical areas are outlined such as "triangle of death" in Campania region (area between Acerra, Nola, and Marigliano municipalities) or the extended area called "Land of Fires" which includes 88 municipalities across Napoli and Caserta provinces [8]. The magnitude of toxic dumping practice is a severe issue for an EU country where statistically all population have access to reliable waste management services. This fact points out that developed countries may have serious gaps in their waste management systems which favor the existence of such wild dumps scattered across rural areas of fly-tipping practices (the USA, Australia, the UK, Mediterranean countries, Central and Eastern Europe). In fact, the "Let's do it! World" movement is a supplementary evidence to this current global environmental issue.

As an example, in the 1990s, in rural Greece there was estimated over 3500 such sites where wastes were illegally disposed without any further treatment (natural depressions, old quarries, gullies, or torrents) [9]. By the mid-1990s, the government of Israel started to replace all unregulated dumps with a rationalized system of large-scale regional landfills [10]. Same threats occurred in the USA [11], and special waste management actions were necessary for rural and remote communities in Canada [12]. New EU members should close and rehabilitate the rural wild dumps until 16 July 2009; meanwhile, the EU candidate countries are expected to solve the problem of wild dumps across rural communities.

Traditional recovery of household waste at the household level, home composting, and animal feed has diverted a part of biowaste fraction from waste dumping into these applications. The improvement of home composting procedure across rural communities is a cost-efficient and an environmentally friendly solution if it is properly performed avoiding the biowaste losses [13]. Reuse and recycling of various items (glass, plastic bottles, construction material, and metal) at household level also mitigate the potential amounts of waste uncontrolled disposed. Frequently, the rural population of low- and middle-income countries relies on solid fuels (firewood, dung, and crop residues) as the energy source for domestic purposes.

Wood, sawdust, paper, and cardboard fractions are used for direct burning as the heating energy source at household level or animal manure in regions without access to forest areas (e.g., high plateau).

Unfortunately, in developing countries, the traditional furnaces are primitive mud stoves and ovens that are extremely air polluting and highly energy inefficient [14]. The incomplete combustion of solid biomass or burning at lower temperature than 800°C leads to exposure of particulate matter (PM), carbon monoxide (CO), oxides of nitrogen and oxides of sulfur (SO_x, NO_x), and phosgene, which has been linked to high morbidity and mortality rates across developing countries [15].

Agricultural wastes (e.g., straws, stalks, husks, wood, and sawdust) are often disposed by burning in open fields with exposure to fire hazard. Household waste (biowaste, plastics, textiles, etc.) are also prone to open burning practices. Mixed wastes may contain hazardous items (e-waste, batteries, oils, solvents, paints, contaminated wood, and pharmaceutical products) which are released into the atmosphere, soil, and groundwaters. The common hazardous substance used in the rural area includes insecticide, pesticide, fungicide, herbicide, chemical fertilizers, chemicals used for fumigation, cleaning agents used in animal husbandry, and medical waste [16]. Such hazardous fraction must be separated, collected, and managed from common household waste.

In worst-case scenario, rural households may have no access to basic utilities (improved drinking water source, sanitation, waste management services), and the near water bodies are polluted by waste dumping and open defecation. In developing countries, especially in rural areas of Africa, India, and China, human waste disposal is a major concern besides household and agricultural waste [17].

There are major gaps in waste collection coverage between larger cities and rural regions across developing and transition countries. A recent study estimates that 1.9 billion people lack waste collection services in rural areas and coverage rate of rural population is under 50% in 105 countries [18]. The amounts of municipal waste generated and uncollected by waste operators or public sanitation services are susceptible to be burnt or uncontrolled dumped, polluting the local environment and threatening the public health. Such wastes pollute the tributaries and rivers, lakes, and coastal areas; thus, floating debris invade marine and ocean ecosystems. Plastic pollution in particular non-compostable microplastics is a notorious threat to marine wildlife, and large areas of oceans called "gyres" concentrate such plastic debris due to the currents (e.g., North Pacific Gyre).

Rural regions without access to formal waste collection services must be encouraged to practice home composting or vermicomposting in order to obtain a qualitative natural fertilizer. Organic farming seeks to reduce external cost, produce good yields, save energy, maintain biodiversity, and keep soil healthy [14]. Composting process may cover various biowaste sources (municipal, sewage, and agricultural) diverting such fractions from open dumping or open burning practices.

If all global domestic wastes derived from organic materials that every year leave the croplands (6.8 billion tons) would be treated by the anaerobic/aerobic process, it could be produced about 4 billion tons of very good soil, avoiding the emissions of 1.4 billion tons of CO_2 eq [19].

Sparsely rural areas which are remote from major urban areas are usually the most neglected by waste management services. Waste operators avoid such areas, and local authorities provide no or low financial resources to provide appropriate public services. In addition, the geographical constraints (mountains, hills, high plateaus, karst regions, and wetlands) makes more difficult to implement proper waste management facilities.

The four cornerstone technologies for agricultural waste and organic fraction of municipal solid waste (OFMSW) suitable for rural communities are animal fodder, briquetting, anaerobic digestion (biogas), and composting with other recycling techniques for solid wastes [14]. Such facilities may serve rural communities without access to formal waste management systems specific to urban areas. These technologies may be integrated into one rural waste complex in order to achieve a desirable zero waste and pollution target [14]. Small anaerobic digesters which use agricultural and food waste may be operational at household level in order to obtain energy (biogas) for cooking and other basic needs. Materials of construction and the design of such digesters are varied based on the geographical location, availability of substrate, and climatic conditions [20]. Thus, in China there are more than 30 million household digesters, India there are 3.8 million, followed by Vietnam with more than 0.5, and Nepal 0.2 million and Bangladesh with 60,000 digesters, while farm-scale digesters are expanding in Europe, the USA, and Canada [21]. Despite the African countries made recent progress on the field where 2619 domestic digesters were installed in 2012 [22], such facilities are still poor exploited due to less availability of technical and operational support, poor digester designs, maintenance, planning, monitoring, lack of awareness, and inadequate dissemination strategy [23]. The common designs include fixed dome (widespread in China), floating drum (widespread in India), and plug flow type (the USA, Peru, etc.) followed by other derivates [20].

In many cases, animal manure, agricultural plant residues (straw, garden wastes, roadside grass), and food waste (OFMSW) are co-digested together to achieve a better nutrient balance in anaerobic digestion process [24]. Community-type biogas digesters have larger volume, and they can produce biogas for several homes instead of one household. Furthermore, public toilets are connected to biogas digesters in India and Nepal [20]. Decentralized facilities are suitable in remote rural regions from which may benefit both industrialized and developing countries. Various geographical regions may provide different biowaste fractions as feedstock for anaerobic digestion process as shown in Nigeria [23].

Biowaste treated in a household biogas digester provides energy for cooking, lighting, and heating along with an improved organic fertilizer in the digest for farmers [20]. The subsidies from the government or local authorities could expand the use of household biogas digesters across rural communities reducing the landfill of biowaste, thus mitigating the Greenhouse gases and leachate emissions into the environment. Developing a user-friendly technology and making it economically viable will enhance the use of biogas digesters which are a boon to low-income and rural people [25].

Large and expensive anaerobic digestion plants and central composting facilities are encountered in regional integrated municipal waste management systems of developed countries which cover cities and surrounding rural areas. Biogas technology is a proven and established technology in many parts of the world such as Germany, the UK, Switzerland, France, Austria, the Netherlands, Sweden, Denmark, Norway, Republic of Korea, Finland, Republic of Ireland, Brazil, China, and India [23].

The European Union imposes that every member state must to reach a 20% share of renewable energies in the total energy consumption by 2020 and to reduce the amount of biodegradable municipal waste that they landfill to 35% of 1995 levels by 2016 (for some countries by 2020) under the Landfill Directive (1999/31/EC). In this context, anaerobic digestion plants could emerge in following years across Europe as alternative energy source to fossil fuels encouraging the transition toward a circular economy approach.

Centralized composting plants usually have as main feedstock the OFMSW of urban areas. However, metropolitan and surrounding rural areas may also contribute with significant amounts of OFMSW in the case of a widespread source-separation collection schemes. The population must be aware that a clean source-separate of biowaste and dry recyclables will improve composting and recycling activities. Intermunicipal cooperation between cities and rural municipalities is mandatory for a successful regional waste management system.

Low technological composting plants should be implemented in rural areas, while in high-density areas, combined anaerobic and aerobic plants with mechanical pretreatment (MBT plants) are preferable due to higher impurities of OFMSW [26].

Waste transportation from source generation (villages) to treatment facilities (transfer station, recycling centers, composting plants, waste to energy plants, and landfills) is a key logistic issue across rural regions.

The budgets of local authorities allocated for waste management sector are limited. Waste management associations group several municipalities or even an entire county/region in order to economically sustain the waste management services.

Major investments are required in order to expand the waste management services from larger cities toward towns and rural localities. EU funds plays an important role in this matter in the case of Central and Eastern European Countries. EU landfill Directive imposes all member states to close the non-compliant urban landfills and rural wild dumps. These are being replaced at the county level by transfer stations, waste to energy plants, or regional sanitary landfills. On the same sites, sorting stations, composting facilities, and crushing plants (construction and demolition waste) may be operational in order to optimize the costs. These integrated waste management systems are based on separate waste collection schemes ("door to door," collection points, and civic amenity sites).

Mixed waste collection must be replaced by such facilities in order to achieve a high rate of waste diversion from landfill sites.

There are two main routes which can help worldwide rural communities to achieve a sustainable waste management system as shown in **Figure 1**. Both routes can be applied at regional level taking into account the specific geographical conditions (natural and socio-economic) which may vary at different scales (village, municipality, county, region, and country).

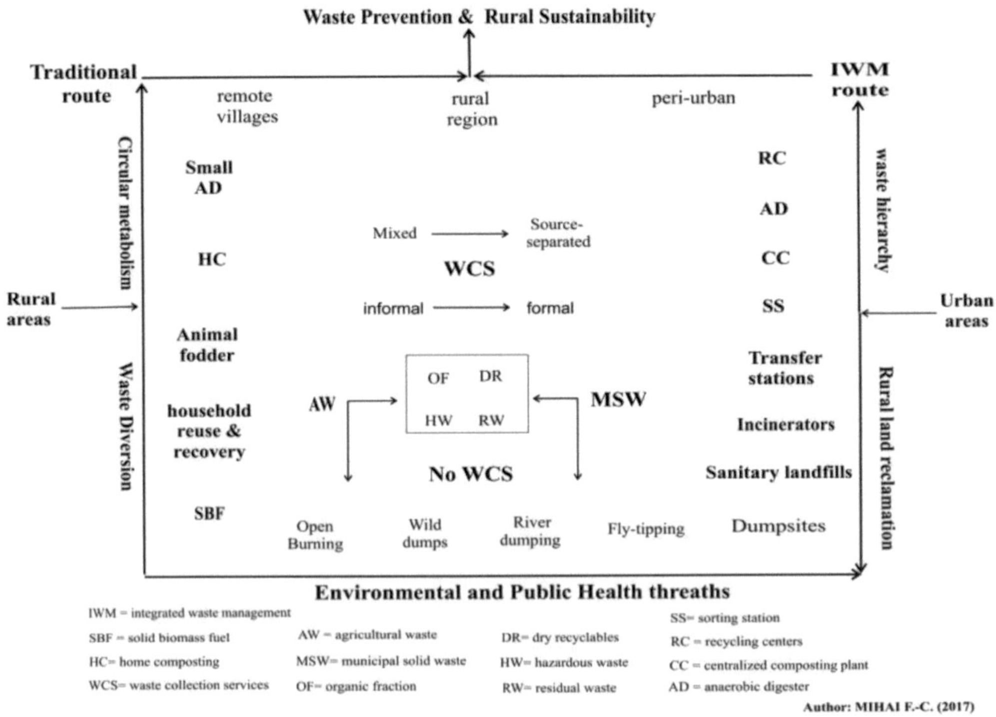

Figure 1. Routes toward waste prevention and rural sustainability.

The rural waste management must rely on a systemic approach involving technical, financial, social, cultural, environmental, and governance aspects. Developing and transition countries must promote smart traditional ways to recycle, reuse, and compost/digest the municipal and agricultural wastes from remote rural regions in order to increase the waste diversion rate from uncontrolled waste disposal practices (open burning, wild dumps, and river/marine dumping).

Generally, rural areas of high-income countries (HIC) are full covered by waste management services in contrast with upper-middle-income countries (UMIC) where the rural population is partially served or low-income countries (LIC) where such services are poor or nonexistent.

In developing countries, informal sector plays a crucial role in diverting recyclables from waste dumping and to provide basic waste collection services, but it is mainly developed in urban and peri-urban areas. Local authorities from many Asian countries operate under severe constraints such as endemic persistence of poverty, unemployment, and underdevelopment which lead to a large informal sector [10]. Animal-driven carts, tricycles, and tractor trailers are frequently used for the transportation of waste across rural communities. The waste management infrastructure is rudimentary; the amounts of waste collected are frequently disposed on open dumps or river banks.

The costs of waste management activities are a heavily burden for small cities and rural localities of developing countries. Such areas are facing a cruel poverty which encourages migration of inhabitants toward urban areas with hope for a better life. Unfortunately, the rapid migration leads to the development of slum areas with the severe challenges in terms of sanitation and waste collection services. On the other hand, urban residents perceive rural areas as sources of raw materials or as places where the most polluting productive activities belong [27]. Environmental injustice operates toward rural areas where urban waste is disposed through large dumpsites, landfills, incinerators, or land application of sludge from urban wastewater [28].

Environmental pollution only seems to be dissipated across sparsely rural regions, but the threats remain at the same level as for urban areas. Furthermore, the pollution activities that occurred in rural areas are more predisposed to be made in an uncontrolled manner. The poor monitoring process and law enforcement lead rural areas to be vulnerable to such practices in both developed or emerging economies.

Home composting and biogas production via home or community digesters are suitable alternatives for rural communities across developing and transition countries where the share of biowaste in the total municipal solid waste fraction is significant and agriculture plays a key role in their economy. However, these practices must be properly performed at the local scale in order to achieve a viable solution for energy and fertilizer demands. Environmental awareness and proper training are crucial to being further developed via governmental programs, local authorities, and civil society. Local municipalities must be supported by financial instruments (subsidies, soft loans, tax incentives, national and international funds) to provide proper facilities for biowaste management.

The regionalization process of waste management infrastructure aims to mitigate the environmental pollution and to expand standardized waste management services across towns and rural municipalities. However, the bureaucracy and delays in the construction process of waste management facilities may lead to serious problems at regional level [29].

Rural-urban relations must be integrated into a sustainable cohesion policy concerning public utilities with a special focus on solid waste management sector.

2. Conclusions

This book intends to draw attention to solid waste management sector toward rural areas where bad practices and public health threats could be avoided through traditional and integrated waste management routes. The expansion of waste collection services across rural municipalities should be a priority for many countries. Agricultural and municipal waste diversion from wild dumps and open burning practices must be avoided through smart solutions at the local level which are cost-efficient particularly in developing countries. The book further examines, on the one hand, the main challenges in the development of reliable waste management practices across rural regions and, on the other hand, the concrete solutions and the new opportunities across the world in dealing with rural solid waste.

Author details

Florin-Constantin Mihai[1]* and Mohammad J. Taherzadeh[2]

*Address all correspondence to: mihai.florinconstantin@gmail.com

1 Department of Research, Faculty of Geography and Geology, "Alexandru Ioan Cuza" University of Iasi, Iasi, Romania

2 Swedish Centre for Resource Recovery, University of Borås, Borås, Sweden

References

[1] Bolton K, De Mena B, Schories G. Sustainable management of solid waste. In: Taherzadeh MJ, Richards T, editors. Resource Recovery to Approach Zero Municipal Waste. CRC Press; USA. 2016. pp 23-40

[2] Taherzadeh MJ, Rajendran K. Factors affecting the development of waste management. Experiences from different cultures. In: Ekström KM, editor. Waste Management and Sustainable Consumption: Reflections on Consumer Waste. Routledge: Earthscan; 2015. pp. 67-88

[3] Stanisavljević N, Ubavin D, Batinić B, Fellner J, Vujić G. Methane emissions from landfills in Serbia and potential mitigation strategies: A case study. Waste Management and Research. 2012;**30**(10):1095-1103

[4] Mihai FC. Spatial distribution of rural dumpsites parameters in Romania. Bollettino dell'Associazione Italiana di Cartografia. 2015;**154**:90-98. DOI: 10.13137/2282-472X/11830

[5] Mazza A, Piscitelli P, Neglia C, Rosa GD, Iannuzzi L. Illegal dumping of toxic waste and its effect on human health in Campania, Italy. International Journal of Environmental Research and Public Health. 2015;**12**(6):6818-6831

[6] Glanville K, Chang HC. Mapping illegal domestic waste disposal potential to support waste management efforts in Queensland, Australia. International Journal of Geographical Information Science. 2015;**29**(6):1042. DOI: 10.1080/13658816.2015.1008002

[7] Kim G, Chang Y, Kelleher D. Unit pricing of municipal solid waste and illegal dumping: An empirical analysis of Korean experience. Environmental Economics and Policy Studies. 2008;**9**:167-176. DOI: 10.1007/BF03353988

[8] Triassi M, Alfano R, Illario M, Nardone A, Caporale O, Montuori P. Environmental pollution from illegal waste disposal and health effects: A review on the "Triangle of Death". International Journal of Environmental Research and Public Health. 2015;**12**(2):1216-1236

[9] Andreadakis AD, Razis Y, Hadjibiros K, Christoulas DG. Municipal solid waste management in Greece. In: Buclet N, Godard O, editors. Municipal Waste Management in Europe A Comparative Study in Building Regimes. Vol. 10. Environment & Management Series. Springer; Springer Netherlands, 2000. DOI: 10.1007/978-94-015-9476-9

[10] Davies AR. The Geographies of Garbage Governance. Interventions, Interactions and Outcomes. Ashgate e-Book; United Kingdom, 2008

[11] Johnson JR. Rural waste management through resource conservation. Bulletin of Science, Technology and Society. 1990;**10**:146-150

[12] UMA Environmental. Small Scale Waste Management Models for Rural, Remote and Isolated Communities in Canada. Prepared for the Canadian Council of Ministers of the Environment Solid Waste Management Task Group. Canada, 1995

[13] Mihai FC, Ingrao C. Assessment of biowaste losses through unsound waste management practices in rural areas and the role of home composting. Journal of Cleaner Production. 2016. DOI: 10.1016/j.jclepro.2016.10.163. In Press

[14] El-Haggar S. Sustainable Industrial Design and Waste Management. Cradle to Cradle for Sustainable Development. Academic Press; USA, 2007

[15] Sidhu MK, Ravindra K, Mor S, John S. Household air pollution from various types of rural kitchens and its exposure assessment. Science of the Total Environment. 2017;**586**:419-429

[16] Chandrappa R, Das DB. Solid Waste Management. Principles and Practice. Springer; (Springer Berlin Heidelberg), Germany, 2012. DOI: 10.1007/978-3-642-28681-0

[17] Epstein E. Disposal and Management of Solid Waste. Pathogens and Diseases. CRC Press; USA, 2015

[18] Mihai FC. One global map but different worlds: Worldwide survey of human access to basic utilities. Human Ecology. 2017;**45**(3):425-429. DOI: 10.1007/s10745-017-9904-7

[19] Masullo A. Organic wastes management in a circular economy approach: Rebuilding the link between urban and rural areas. Ecological Engineering. 2017;**101**:84-90

[20] Rajendran K, Aslanzadeh S, Taherzadeh MJ. Household biogas digesters—A review. Energies. 2012;**5**:2911-2942. DOI: 10.3390/en5082911

[21] Kabir MM, Forgács G, Taherzadeh MJ, Horváth IS. Biogas from wastes: Processes and applications. In: Taherzadeh MJ, Richards T, editors. Resource Recovery to Approach Zero Municipal Waste. CRC Press; USA, 2016. pp.107-140

[22] Surendra KC, Takara D, Hashimoto AG, Khanal SK. Biogas as a sustainable energy source for developing countries: Opportunities and challenges. Renewable and Sustainable Energy Reviews. 2014;**31**(846):859

[23] Akinbomi J, Brandberg T, Sanni SA, Taherzadeh MJ. Development and dissemination strategies for accelerating biogas production in Nigeria. BioResources. 2014;**9**(3):5707-5737

[24] Nayal FS, Mammadov A, Ciliz N. Environmental assessment of energy generation from agricultural and farm waste through anaerobic digestion. Journal of Environmental Management. 2016;**184**:389-399

[25] Rajendran K, Aslanzadeh S, Johansson F, Taherzadeh MJ. Experimental and economical evaluation of a novel biogas digester. Energy Conversion and Management. 2013;**74**(1):83-191

[26] Sánchez A, Gabarrell X, Artola A, Barrena R, Colón J, Font X, Komilis D. Composting of wastes. In: Taherzadeh MJ, Richards T, editors. Resource Recovery to Approach Zero Municipal Waste. CRC Press; USA, 2016. pp 77-116

[27] Gallaud D, Laperche B. Circular Economy, Industrial Ecology and Short Supply Chain. Vol. 4. Smart Innovation Set. John Wiley & Sons, Inc; USA, 2016

[28] Kelly-Reif K, Wing S. Urban-rural exploitation: An underappreciated dimension of environmental injustice. Journal of Rural Studies. 2016;**47**:350-358. http://dx.doi.org/10.1016/j.jrurstud.2016.03.010

[29] Mihai FC. Waste collection in rural communities: Challenges under EU regulations. A case study of Neamt County, Romania. Journal of Material Cycles and Waste Management. 2017. DOI: 10.1007/s10163-017-0637-x

Chapter 2

Household's Willingness to Accept Waste Separation for Improvement of Rural Waste Bank's Effectivity

Christia Meidiana, Harnenti Afni Yakin and
Wawargita Permata Wijayanti

Additional information is available at the end of the chapter

http://dx.doi.org/10.5772/intechopen.69428

Abstract

Waste Bank, a form of public-community participation (PCP) system in managing the households' solid waste problems, becomes popular in Indonesia. Waste Bank program involves community and provision of incentives to them and requires public acceptance measured through willingness to accept (WTA). Therefore, this study aims to estimate households' WTA compensation in terms of inorganic waste separation adopting the contingent valuation method. It measures also the effectiveness of waste bank (WB) and community adaptability on WB in Gili Trawangan Island (GTI), Indonesia. The community acceptance is measured using Willingness to Accept (WTA) the obligation to separate waste. Fully structured questionnaires are filled in by 94 respondents through random sampling to evaluate the current WB. The result shows that the score for overall equipment effectiveness (OEE), adaptability and acceptance of waste bank is 12.67%, 1.50, and 37.5% respectively. It indicates that waste bank is relatively difficult to be developed, people and waste institution has low adaptability with current waste bank system and only some people want to participate in waste bank. Based on this result, WTA is measured to determine the optimum price of recyclable waste sold to waste bank to improve the WB's performance and to increase community acceptance.

Keywords: waste bank, willingness to accept, overall equipment effectiveness

1. Introduction

Rural solid waste (RSW) has less priority in most of the developing countries [1]. Urbanization and the fast population growth in urban area have come to local authorities' attention in all sectors including municipal solid waste. RSW should be part of integrated solid waste management since the waste in rural areas increases in quality and quantity because of the lifestyle change and income increase. Solid waste management (SWM) requires a systematic approach which integrates environmental effectiveness, public acceptance, and economic affordability [2]. Public acceptance refers to the favorable reception and the active approval and adoption of newly introduced technical devices and systems [3]. Public acceptance in waste management can be measured through public participation rate. Public participation is acknowledged as the method to attain sustainable WM, and it can bridge the gap between government and citizens in environmental conflict management [4, 5]. Public participation in solid waste management should be addressed toward the "waste as resource" and the "waste as income generator" in household units [6]. It serves the purpose of daily waste disposal decrease, waste utilization as resources for certain local production, income generator, and benefit agent for the households involved in solid waste management. Households' involvement in solid waste management may be in the form of waste separation and recycling. Waste management (WM) strategies involving waste separation and recycling will only be successful if they are supported by the public including the local residents. Local residents are nonignorable stakeholders in both the daily WM and the decison-making process because they are both the subject and the object of waste management services [7, 8]. The performance patterns and community's attitudes, shaped by the local cultural and social background, determine the structure and functions of public participation [9]. Hence, the challenge for WM is to enhance public participation nowadays. In Indonesia, the number of researches focusing on public willingness to participate in WM and its influencing factors is still low. These factors could be demographic variables, i.e., age, gender, and household typology, knowledge, and recycling time [10–12] as well as educational level, occupation or income level [13–17]. The findings of each study often depend on the sample used. Identifying these factors and their importance may be beneficial for the improvement of public participation in WM since it depends on local situation. Design of a successful scheme may not necessarily be replicable elsewhere [18]. Public acceptance can be reflected by the willingness to accept (WTA).

The contingent valuation method (CVM) was applied in this study to draw people's willingness to accept (WTA) economic sacrifices to separate waste. The contingent valuation method (CVM) was claimed to be the most suitable tool available to measure nonmarket value. Previous studies used it to measure public goods and services [19] and to assess farmers' participation preference [20]. Properly designed willingness to accept (WTA) can estimate the strength of demand for who are willing or never willing to consume a certain good [21].

WTA waste separation of households residing in Gili Trawangan was measured, and the expected compensation for it was assessed by asking the households for their WTA. Gili Trawangan is a famous tourist destination island. Every year, there is 11.8% visit increase to the island leading to waste production increase. The main sources of waste in this island are households, hotels, and restaurants accounting for 602 ton/day of waste, out of which about 42% is inorganic. Currently, there is no waste management in the island provided by the local government. There are community initiatives that conduct waste separation and waste bank to reduce inorganic waste, i.e., plastics, paper, metal, and glasses, and to bring income by selling it. Unfortunately, public participation in waste separation is very low which may be caused by the ineffectiveness of the waste bank. Therefore, this research aims to measure the effectiveness of waste bank, public adaptability, and public acceptance in environmental improvement through waste separation and waste bank.

This chapter is divided into three main parts. The first part explains the methodology applied. The second part outlines the result of village identification, data collection, and data analysis. This section is followed by the measurement of waste bank effectiveness, public adaptability, and willingness to accept (WTA) waste separation. The last section is conclusions explaining about the findings and the recommendations for waste management improvement in Gili Trawangan Island (GTI).

2. Research method

The area of study is located in Gili Indah Village, Gili Trawangan Island (GTI) Lombok Utara Regency, Nusa Tenggara Barat Province, Indonesia (**Figure 1**). The area belongs to one of the strategic development zones in Nusa Tenggara Barat Province. Tourism sector in GTI contributes 60–70% to the total income of local government [22]. Rapid increase of visit in GTI leads to more waste volume. In 2015, Community forum on Environment measured that the average waste generation in GTI is 17 ton/day where 6.2 ton is inorganic waste. Currently, inorganic waste is managed by WB Bintang Sejahtera. However, WB's performance is relatively low since the amount of inorganic waste that can be treated through this WB is still low. Based on the population in GTI, samples were determined using stratified random sampling. Unit analysis of the study was household. Eighty households were selected as respondents and they were provided with questionnaires to gain required data for measuring the willingness to accept (WTA). Bidding game format was used to assess the WTA of households. Waste bank effectiveness is measured using Eq. (1) which is equation of overall equipment effectiveness (OEE); A, P, and Q represent availability, performance, and quality, respectively. Each variable is calculated using Eq. (2), Eq. (3), and Eq. (4).

$$OEE = A \times P \times Q \tag{1}$$

$$A = \frac{Aa}{Ra} \times 100\% \tag{2}$$

$$P = \frac{W_i \times Tq}{Aa} \times 100\% \tag{3}$$

$$Q = \frac{Aa}{Tq} \times 100\% \tag{4}$$

The effectiveness of waste bank is scored based on the percentage gained from the calculation as shown in **Table 1**.

Analysis on public adaptability was conducted afterward to find out whether the related stakeholders (community and institutions) can accept the continuation of waste bank program. Some indicators were introduced and scoring was given for each indicator ranging from 0 to 4. The result was used as a reference to scale public adaptability on waste bank program. **Table 2** shows the adaptability level based on the score.

Furthermore, willingness to accept (WTA) of the community to separate waste and sell it to waste bank was measured. Bidding game was used to get the optimum price for recyclable materials sold to the waste bank. Bidding game provides flexibility to the respondent for giving answer without losing the context since the lowest value is determined beforehand.

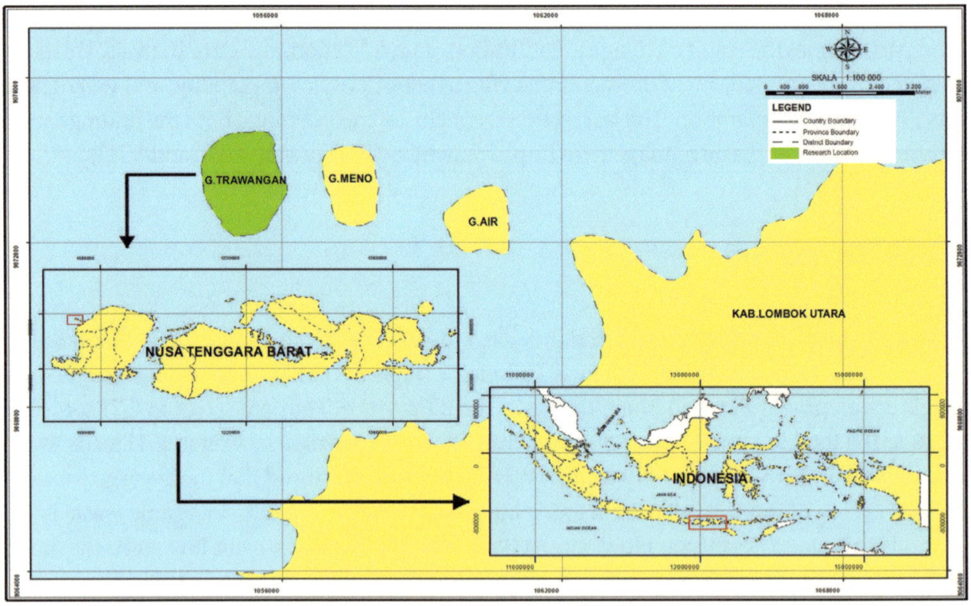

Figure 1. Research location.

Percentage of OEE	Criteria	Score
If OEE = 100%	• Waste bank is perfectly run • Produces only programs with significant outcomes • Fast service and no *downtime*	4
If 85% ≤ OEE <100%	• Waste bank is run optimally but can be more improved • Produces some program and most of them is implemented • Long-term goal: goal-oriented programs	3
If 60% ≤ OEE <85%	• Waste bank is fairly good • Produce some programs and some have not been implemented • Wide opportunity for more *improvement*	2
If 40% ≤ OEE <60%	• Waste bank is average • Produce some programs and only few have not been implemented • Frequent *downtime*	1
If OEE < 40%	• Waste bank is poor • Hard to be improved • Most of the programs are not implemented • Required deep observation to find out the reasons for the poor condition	0

Table 1. Criteria for measuring the OEE.

Scores	Remarks
<1.00	Not capable to adapt
$1.00 \leq x < 2.00$	Less capable to adapt
$2.00 \leq x < 3.00$	Adequately capable to adapt
$3.00 \leq x < 4.00$	Capable to adapt
4.00	More capable to adapt

Table 2. Adaptability level.

3. Result and discussion

3.1. Waste generation and composition

Waste sources in GTI are mainly households (HH) and hotels (HT) generating waste of 20–30 and 100–300 kg/day, respectively. The compositions of organic and inorganic wastes are 65 and 35%, respectively. Totally, about 17.72 ton waste is generated in GTI per day as shown in **Table 3**.

Inorganic waste is mainly comprised of plastic, glass bottles, food wrap, and tin which comes from commercial facilities, i.e., restaurants, hotels, guest houses, bars, and recreation areas. Some of these wastes have been managed by Bintang Sejahtera WB established in 2015.

3.2. Waste bank in GTI

Bintang Sejahtera WB is a community-based waste management system that aims to reduce waste and to get benefit from waste. It accepts inorganic waste separated by the households including plastic bottle/glass, aluminum tin, plastic bag, paper, and cardboard. The condition of Bintang Sejahtera is shown in **Figure 2**.

In 2016, the daily separation rate of Bintang Sejahtera WB was 4.430 ton/day or 25% of total waste generation in GTI in which 3.145 ton was plastic. The waste was sold to some industries in other cities outside the island with the price ranging from Rp 200 to 9000. The selling price for each waste type is shown in **Table 4**.

Location	Waste types	Waste generation (ton/day)	Average waste generation (ton/day/person)
Gili Trawangan	Organic	11.52	0.005
	Inorganic	6.20	0.002
	Total	17.72	0.007

Table 3. Waste generation in GTI (2015).

Figure 2. Condition of Bintang Sejahtera WB.

No	Waste types	Price (USD)
1	Plastic bottle	0.152
2	Plastic glass	0.152
3	Aluminum tin	0.682
4	Cardboard and paper	0.076
5	Plastic bag	0.015
6	Tetrapack	0.023

*One rupiah equals USD 13.198 based on rate from Indonesian Central Bank.

Table 4. Selling price of waste in Bintang Sejahtera WB in 2016.

There are several activities that are conducted every day, such as collecting waste from households, restaurants, bars, and others. Then, WB staffs sort the organic and inorganic wastes, weigh them (**Figure 3**), and record it (**Figure 4**). The organic waste will be used to make a natural fertilizer by the environmental community initiative staffs. Meanwhile, the inorganic waste will be recycled or reused.

Bintang Sejahtera WB addresses not only profit but also social development and environmental improvement. Through waste bank, villager's welfare can be increased though better income and healthier environment. Some programs are offered by Bintang Sejahtera WB, such as health savings, education savings, and electricity and water savings, which can be claimed by the vil-

Figure 3. Weighing the waste in Bintang Sejahtera WB.

lager as a member of WB when it is needed. Bintang Sejahtera WB has cooperation with the environmental community initiative to collect waste from beaches and with the local government to provide collection system to transport the waste. It also offers seminars and trainings for local people in terms of waste treatment (composting and reuse-reduce-recycle method).

3.3. Waste bank effectiveness

Waste bank is an implementation of Reuse, Reduce, and Recycle (3R) of inoragnic waste in GTI. However, there is no evaluation of WB effectiveness so far. Therefore, the evaluation is conducted to measure the level of effectiveness based on three subjects that is availability, performance, and quality.

3.3.1. Availability

Availability is defined as the ability of WB to conduct activities related to waste management within a certain time; it refers to operational time. The ability is the ratio of existing operational time to planned operational time. Currently, the operational time of WB is 8 h in compliance with planned operational time. It indicates that the availability of WB is in proper condition.

3.3.2. Performance

Performance is the achievement of WB in a certain period based on the existing operational time, ideal time allocated for each activity, and the number of WB's activities in a certain

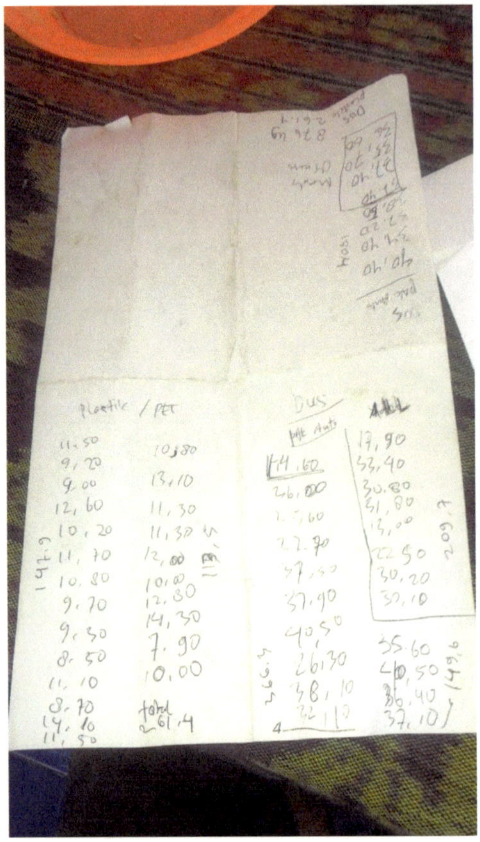

Figure 4. Waste record and list in Bintang Sejahtera WB.

period. The operational time of WB is 8 h accommodating four activities, i.e., waste separation, waste compacting, waste weighing, and data recording. The time allocated for each activity is 4 h, 2 h, 15 min, and 5 min for separation, waste compacting, waste weighing, and data recording, respectively, and an additional 1 h for lunch break.

3.3.3. *Quality*

WB quality is determined by analyzing the WB program's success in its implementation and its significant contribution to benefit the community. WB has a good quality when the above criteria are fulfilled. The quality is measured based on the number of WB's program which has been implemented. Bintang Sejahtera WB has six programs where five are savings for health, education, holiday, electricity, and water and one is for environmental hygiene and conservation. Calculation of WB effectiveness using Eq. (5) is shown in the **Table 5**.

$$OEE = Availability * Performance * Quality = 1 * 0.79 * 0.16 = 0.1267 = 12.67\% \quad (5)$$

Variables	Indicators	Results	Notes
Availability	• Current operational time of WB (A) = 8 h = 480 min • Planned the operational time allocated for running the WB (R) = 8 h = 480 min	$A = \frac{A}{Ra} \times 100\,\% = \frac{480}{480} \times 100\,\%$ $= 100\,\% = 1$	Availability of WB is maximum since operational time fulfill planned time allocation (8 h)
Performance	• Current operational time of WB (A) = 8 h = 480 min • Number of WB's activities (N) = 4 i.e., waste separation, compacting, weighing, and recording. • Ideal operational time for each activity (W_i), i.e., separation, compacting, weighing, and recording for 240, 120, 15, and 5 min, respectively.	$P = \frac{W_i \times N}{A}$ $= \frac{((1 \times 240) + (1 \times 120) + (1 \times 150) + (1 \times 5))}{480}$ $= 79\,\% = 0.79$	Performance Bintang Sejahtera WB is not maximum. There is abandoned 40 min from total 8 h operational time.
Quality	• Number of program implemented (Aq)= 1 program • Number of available total program (Tq) = 6 program	$Q = \frac{Aq}{Tq} \times 100\,\% = \frac{1}{6} \times 100\,\%$ $= 16\,\% = 0.16$	Programs offered by Bintang Sejahtera WB is not maximum. Only one program is implemented caused by the public participation

Table 5. Calculation of WB's effectiveness in GTI.

Multiplying three variables come to the result that OEE is 12.67%. This value is below 40%. Referring to **Table 1**, WB has zero score indicating that waste bank has poor effectiveness and is hard to be improved. Improvement is required to pace waste generation increase in GTI projected to be 23.23 ton/day in 2020 where 35% of it is inorganic waste. Otherwise, GTI will face waste problems because landfill in GTI is approaching its maximum capacity.

Analyzing the OEE, it can be recognized that low OEE value is caused by low quality value of WB. Low quality value is determined by the number of implemented programs which is only one from six programs offered. Low public participation is the reason for this. Waste separation is not common for the villager in GTI, and only small number of HHs is involved in waste separation. Thus, the number of WB customer is also very low. Furthermore, WB's performance is not maximum because there is 40 min remaining time unused for waste management activities.

Improvement of WB' effectiveness may increase public participation which requires public adaptability to WB's program in GTI. Therefore, public adaptability is necessary to be measured. Evaluation of public adaptability in GTI may contribute to find out the adaptability level, its factors, and the possible solutions.

3.4. Public adaptability

Public adaptability to WB is defined as community's and institution's adaptability for being active in WB program and is assessed based on reason/motivation and behavior [23]. Community refers to the villager of GTI, while institution refers to the Bintang Sejahtera WB, the environmental community initiative, and the local government.

3.4.1. Community's motivation and behavior

Community's motivation and behavior is a push factor for the villager in GTI to participate in WB's programs. Survey results showed some reasons for motivation to be engaged or not in WB's program, i.e., 53.8% villagers had no motivation to be active in the programs because of nescience of WB's purposes and benefits and subsequence of WB's program; 42.8% villagers were motivated to be active in which 30.0, 8.8, and 7.5% villagers had both environmental awareness and additional income, only environmental awareness, and only income addition, respectively. The percentage affirms the behavior of the community where 83.8% villager do not separate waste currently.

3.4.2. WB staff's motivation and attitude

WB Staffs have an important role in WB implementation. There are six persons managing the process in WB comprising waste separation, compacting, weighing, and data recording. Their motivation may be the factor influencing WB's effectivity. The result shows that 50% staff has motivation to be involved in WB for environmental awareness and the rest is for additional income.

3.4.3. Community initiative staff motivation and attitude

The environmental community initiative staffs support the WB in waste transportation from waste sources (HHs and commercial facilities) to WB and composting center. All staffs have high motivation and their behavior reflect high commitment to improve waste management in GTI. They also plan to develop organic farming in GTI within 2 years.

3.4.4. Local government officer's motivation and attitude

Some related local planning has been set including transfer point construction, an incinerator erection, and vehicle procurement.

The analysis comes to the result that each stakeholder has different adaptability level. **Table 6** describes the adaptability level of stakeholders of WB in GTI.

It can be summed up that the average adaptability score is 1.80. Referring to **Table 6**, the score indicates that has less capability to adapt WB because the score lies between 1.00 and 2.00.

3.5. Willingness to accept

WTA of HHs is measured to determine the expected compensation to separate waste and sell it to the WB. Furthermore, WTA may reflect the public through eliciting questions in

Stakeholders	Motivation and attitude	Score
Community	Less motivation of GTI community makes most of them not to support WB activities	2.00
WB staff	All of WB staff have been motivated due to economic and environmental added values of WB	3.00
Community initiative staffs	Most of their programs have been well conducted (two out of three programs)	3.00
Local government officers	They only conducted one out of three programs	1.00
Average score		1.80

Table 6. Bintang Sejahtera WB adaptability.

questionnaires. Villager who accepts the WB program is asked further for acceptable price for the waste transported to WB. **Table 7** explains the acceptable price for each waste type for 94 respondents representing the whole HHs in GTI.

Aluminum tin has the highest and plastic bag has the lowest acceptable prices compared to other waste types. Furthermore, a comparison between the acceptable price and the current market price for the waste set by the middleman is conducted to find out whether the price is reasonable to be set or not. It is expected that public participation in WB

Waste types	Expected waste price by community (Rp/kg)		Most acceptable price (Rp)
Plastic bottle	Most expensive	0.189–0.265	0.227
	Cheapest	0.038–0.114	0.114
Glass bottle	Most expensive	0.189–0.265	0.227
	Cheapest	0.076–0.129	0.114
Small beer bottle	Most expensive	0.023–0.045	0.038
	Cheapest	0.008–0.015	0.008
Big beer bottle	Most expensive	0.076–0.114	0.114
	Cheapest	0.023–0.038	0.038
Ketchup bottle	Most Expensive	0.061–0.114	0.076
	Cheapest	0.008–0.038	0.023
Aluminum tin	Most expensive	0.833–1.137	0.985
	Cheapest	0.227–0.492	0.379
Cardboard and paper	Most expensive	0.114–0.189	0.152
	Cheapest	0.038–0.076	0.076
Plastic bag	Most expensive	0.023–0.053	0.038
	Cheapest	0.008–0.015	0.008
Tetrapack	Most expensive	0.038–0.053	0.045
	Cheapest	0.008–0.023	0.008

*One rupiah equals USD 13.198 based on rate from Indonesian Central Bank.

Table 7. Acceptable waste selling price in GTI.

Waste types	Acceptable price (Rp/kg)	WB's selling price (Rp/kg)	Middleman's selling price (Rp/kg)
Plastic bottle	0.114–0.227	0.152	0.152
Plastic glass	0.114–0.227	0.152	0.227
Small beer bottle (glass)	0.008–0.038	0.000	0.023
Big beer bottle (glass)	0.038–0.114	0.000	0.061
Ketchup bottle (glass)	0.023–0.076	0.000	0.045
Aluminum tin	0.379–0.985	0.682	0.758
Cardboard and paper	0.076–0.152	0.076	0.114
Plastic bag	0.008–0.038	0.015	0.000
Tetrapack	0.008–0.045	0.023	0.000

*One rupiah equals USD 13.198 based on rate from Indonesian Central Bank.

Table 8. Waste selling price.

increases when WB offers relatively higher selling price. **Table 8** shows the comparison of waste selling price acceptable for the HHs, set by WB and middleman. It is obvious that WB generally sets lower selling price. Higher selling price offered by the middleman may be an obstacle. Moreover, some waste type such as small beer bottle, big beer bottle, and aluminun tin are not accepted by WB although the generation of these waste types is relatively high.

Acceptable waste selling price is within the price range offered by both WB and middleman indicating that most HHs can accept the WB's program. HH's WTA is reasonable to be implemented with the most acceptable price as a compensation for waste separation done by the HHs.

4. Conclusion

The result from the OEE calculation show that:

1. Availability of waste bank is 100% indicating that time provision for service is very good for conformity of the time allocation.
2. Performance of waste bank is 79% indicating that performance is not optimal since there are 40 min remaining from the whole work hours.
3. Quality of waste bank is 16% indicating that the quality is poor caused by low involvement of community and low implementation rate of existing programs (one out of six).
4. OEE is 12.67% which equals to score 0 indicating that waste bank is difficult to be improved.

http://dx.doi.org/10.5772/intechopen.69428

Waste type	Acceptable price for recyclable material (Rp/kg)
Plastic bottle	0.114–0.227
Plastic glass	0.114–0.227
Small beer bottle	0.008–0.038
Big beer bottle	0.038–0.114
Ketchup bottle	0.023–0.076
Aluminum can	0.379–0.985
Cardboard/paper	0.076–0.152
Plastic bags	0.008–0.038
Tetrapack	0.008–0.045

Table 9. WTA for waste separation reflected by optimum price for recyclable waste.

5. The availability score is 1.5 and community acceptance is 37.5%.
6. WTA is reflected by the optimum price accepted by the community as a compensation if they separate waste and sell waste to the waste bank. WTA for waste separation reflected by optimum price for recyclable waste is shown in **Table 9**.

5. Recommendations

There are some recommendations for improvement of WB's effectiveness based on the result of the analysis:

1. Provision of pickup service for members.
2. Employment of remaining 40 min to increase the customer service.
3. Cooperating with owners of commercial facilities to separate waste and providing pickup service.
4. Public dissemination about the WB's benefit through regular open hearing.
5. Increasing waste selling price and expanding acceptable waste type.

Author details

Christia Meidiana*, Harnenti Afni Yakin and Wawargita Permata Wijayanti

*Address all correspondence to: c_meideiana@ub.ac.id

Faculty of Engineering, Universitas Brawijaya, Malang, Indonesia

References

[1] Zarate MA, Slotnick J, Ramos M. Capacity building in rural Guatemala by implementing a solid waste management program. Waste Management. 2008;**28**(12):2542-2551. DOI: 10.1016/j.wasman.2007.10.016

[2] Xiao L, Zhang G, Zhu Y, Lin T. Promoting public participation in household waste management: A survey based method and case study in Xiamen city, China. Journal of Cleaner Production. 2017;**144**:313-322. DOI: 10.1016/j.jclepro.2017.01.022

[3] Gaul S, Ziefle M. Smart home technologies: Insights into generation-specific acceptance motives. In: Holzinger A, Miesenberger K, editors. HCI Usability E-Incl. Heidelberg: Springer; 2009. pp. 312-332. DOI: 10.1007/978-3-642-10308-7_22

[4] Joseph K. Stakeholder participation for sustainable waste management. Habitat International. 2006;**30**(4):863-871. DOI: 10.1016/j.habitatint.2005.09.009

[5] Lin T, Guo XH, Zhao Y, Pan LY, Xiao LS. Study on differences in resident's environmental awareness among various communities in a peri-urban area of Xiamen, China. International Journal of Sustainable Development and World Ecology. 2010;**17**(4):285-291. DOI: 10.1080/13504509.2010.487995

[6] Singhirunnusorn W, Donlakorn K, Kaewhanin W. Contextual factors influencing household recycling behaviours: A case of waste bank project in Mahasarakham municipality. Procedia – Social and Behavioral Sciences. 2012;**36**:688-697. DOI: 10.1016/j.sbspro.2012.03.075

[7] Garnett K, Cooper T. Effective dialogue: Enhanced public engagement as a legitimising tool for municipal waste management decision-making. Waste Management. 2014;**34**(12):2709-2726. DOI: 10.1016/j.wasman.2014.08.011

[8] Tai J, Zhang W, Che Y, Feng D. Municipal solid waste source-separated collection in China: A comparative analysis. Waste Management. 2011;**31**(8):1673-1682. DOI: 10.1016/j.wasman.2011.03.014

[9] Peter S, et al. Conceptual Framework for Municipal Solid Waste Management in Low-income Countries. 1st ed. Washington DC: World Bank; 1996. 59 p

[10] Vicente P, Reis E. Factors influencing households' participation in recycling. Waste Management and Research. 2008;**26**(2):140-146. DOI: 10.1177/0734242X07077371

[11] Cox J, Giorgi S, Sharp V, Strange K, Wilson DC, Blakey N. Household waste prevention – A review of evidence. Waste Management and Research. 2010;**28**(3):193-219. DOI: 10.1177/0734242X10361506

[12] Gellynck X, Jacobsen R, Verhelst P. Identifying the key factors in increasing recycling and reducing residual household waste: A case study of the Flemish region of Belgium. Journal of Environment Management. 2011;**92**(10):2683-2690. DOI: 10.1016/j.jenvman.2011.06.006

[13] Kinnaman TC, Fullerton D. Garbage and recycling with endogenous local policy. Journal of Urban Economics. 2000;**48**(3):419-442. DOI: 10.1006/juec.2000.2174

[14] Barr S, Gilg A, Ford N. Defining the multi-dimensional aspects of household waste management: A study of reported behavior in Devon. Resources, Conservation and Recycling. 2005;**45**(2):172-192. DOI: 10.1016/j.resconrec.2004.12.007

[15] Saphores JM, Nixon H, Ogunseitan OA, Shapiro AA. Household willingness to recycle electronic waste: An application to California. Environmental Behavior. 2006;**38**(2):183-208. DOI: 10.1177/0013916505279045

[16] De Feo G, De Gisi S. Public opinion and awareness toward MSW and separate collection programmes: A sociological procedure for selecting areas and citizens with a low level of knowledge. Waste Management. 2010a;**30**(6):958-976. DOI: 10.1016/j.wasman.2010.02.019

[17] Bringhenti JR, Risso Gunther WM. Social participations in selective collection program of municipal solid waste. Engenharia Sanitaria e Ambiental. 2011;**16**(49):421-430. DOI: 10.1590/S1413-41522011000400014

[18] Tucker P, Grayson J, Speirs D. Integrated effects of a reduction in collection frequency for a curbside newspaper recycling scheme. Resources, Conservation and Recycling. 2000;**31**(2):149-170. DOI: 10.1016/S0921-3449(00)00078-1

[19] Loomis J, Kent P, Strange L, Fausch K, Covich A. Measuring the economic value of restoring ecosystem services in an impaired river basin results from a contingent valuation survey. Ecological Economics. 2000;**33**(1):103-117. DOI: 10.1016/S0921-8009(99)00131-7

[20] Dupraz P, Vermersch D, Defrahan BH, Delvaux L. The environmental supply of farm households – A flexible willingness to accept model. Environmental and Resource Economics. 2003;**25**(2):171-189. DOI: 10.1023/A:1023910720219

[21] Moon W, Rimal A, Balasubramanian SK. Willingness-to-accept and willingness-to-pay. In: Annual Meeting of American Agricultural Economics; 1-2 August 2004; Denver, Colorado. 2004

[22] Ari. 70% Pendapatan Asli Lombok Utara Berasal Dari Sektor Pariwisata [Internet]. 2013. Available from: http://www.radarlombok.co.id/realisasi-pad-sudah-lampaui-target.html [Accessed: 15 December 2015]

[23] Helena K. Model konseptual untuk mengukur adaptabilitas bank sampah. Jurnal Lingkungan. 2014;**IX**(1)

Chapter 3

The Role of the Informal Sector in a Rurbanised Environment

Petra Schneider, Le Hung Anh, Jan Sembera and Rodolfo Silva

Additional information is available at the end of the chapter

http://dx.doi.org/10.5772/intechopen.70169

Abstract

Economic activities performed by rural populations linked to informal trading and markets have not received a broad attention in the literature. Thus, the question of the present investigation is the role of the informal sector in a rurbanised environment, and if there are differences in the waste management activities of the informal sector in cities and in an urbanised rural environment. To obtain information about the informal waste pickers in the rural areas, data were collected directly through a questionnaire from the following countries (sorting in alphabetic order): Austria, the Czech Republic, Germany, Jordan, Mexico, Nepal, South Africa and Vietnam. The methodology used for the data collection consisted of a background analysis (with a literature review), complemented with the collection of empirical evidence, field interviews and partially local field analysis. The informal collection of waste is a phenomenon that results in principle from social differences within society and the population. Therefore, it is not surprising that the perception of the activities of informal waste collectors in the scientific literature refers to developing and emerging countries, since social differences are more pronounced. These informal waste management systems in low- and middle-income countries exist usually in parallel with formal waste management systems, a fact that applies for urban as well as rural areas, and might be considered as a result of rurbanisation. The case studies show the development of the informal sector as an important part of the waste management activities, when a country evolves. With increasing economic development, the importance of the informal sector is shrinking step by step in relation with the improvement of the formal activities. Even this development goes faster in urban areas; the conclusion applies also to rural areas.

Keywords: informal waste collection, informal recycling, waste collection in rural areas

1. Introduction

The term 'rurban' refers to a region which has both urban and rural characteristics. Rurbanisation may be due to either urban expansion or rural migration, leading to urban–rural interactions, which result in an urbanised lifestyle in rural areas. This development manifests in rapid urbanisation of the rural population—lifestyles and mind sets—perceiving cities as a source of income, stability and a possibility for better living conditions. The perception of rurbanisation goes back nearly a century. The term was firstly used by Sorokin & Zimmerman [1]. Also, Parson [2] highlighted the idea of rurbanisation, describing rurban communities as rural socio-geographic spaces where styles of life and the standard of living have changed so much that they resemble those in urban localities. This phenomenon also found in massive migration from rural to urban areas. Later research on rurbanisation by Chapuis & Brossard [3] described a population growth phenomenon observed in the rural environment due to the effect of changing the rural–urban migration patterns from the urban to the rural direction. This phenomenon was named 'rural rebirth', characterised by community policies [4], receptivity, land use [5], utilising neighbours [6], agricultural development [7], tourist sites, secondary residences and available homes [8], as well as endless options [9].

Rural rebirth describes the migratory flow caused by the effects of rurbanisation on rural livelihood [9]. The rural rebirth phenomenon also reflects a special economic situation: the financial potential to afford to live a separated life in the countryside. The definition of rurbanisation exposes its effect on rural patterns: '*rurbanisation is a process of altering rural forms with pre-selected urban patterns and lifestyles, which creates new genetically altered rurban forms*' [10, 11]. Nowadays, both types of migration are observed in parallel: rural–urban migration mainly in emerging countries resulting in the formation of megacities, and urban–rural migration mainly in industrial countries. Also, the medial impact forces the urbanisation of rural livelihood through advertising and sales strategies. Rurbanisation leads to a habit change in waste generation: while poor population from rural areas mostly produces organic and fast biodegradable wastes, the more rurbanised population is consuming in a different way, causing a double consumption in comparison to traditional lifestyle and an increased waste generation of plastic, glass, metal and electronics [12]. Recyclable materials are of interest for recyclable waste dealers, leading to the situation that rurbanisation causes activities of informal waste pickers also in the rural area.

The term 'informal' does not give a clear definition in the literature yet. According to Chi et al. [13], informal activities are possible to be carried out '*due to lack of legislation, structure or institutionalisation in a way out of the different levels and mechanisms of the official governmental power*'. Furthermore, they can be characterised as '*not registered, and characterised as illegal*'. Informal actions can therefore not be equated with such illegal acts, since the term 'informal' additionally involves legal grey zones. The term 'informal' thus also includes non-regulated acts and unclear defined rules [14]. The informal sector is characterised by labour-intensive, largely unregulated and unregistered, low-technology manufacturing or provision of waste collection services [15]. Informality is usually associated with undesirable developments such as tax evasion, unregulated enterprises and even environmental degradation [16]. Mainly in low- and middle-income countries, the informal sector especially in the urban area reaches a significant proportion of the waste collection activity in solid waste management (SWM) as reported by Scheinberg et al. [17]: Belo Horizonte, Brazil—6.9%; Canete, Peru—11%; Delhi, India—27%; Dhaka, Bangladesh—18%; Managua, Nicaragua—15%; Moshi, Tanzania—18%; Quezon City, Philippines—31%. For rural areas, information on the percentage of informally collected waste is very rare. Even informal sector entrepreneurs in the past did not pay taxes, not have a trading license and are not included in social welfare or government insurance schemes [18], since a few years there are strong activities in many developing countries to include the informal sector into the official waste management system [19, 20]. This leads to the situation that the informal sector generally achieves high recovery rates (up to 80%) because the ability to recycle is vital for the livelihood of people involved [19, 20].

The official waste management system in urban and urbanised areas could not be managed without waste pickers, scrap collectors, traders and recyclers. Although not officially recognised, they often perform a significant percentage of waste collection services, in many cases at no cost to local authorities, central governments or residents. By its nature, the activity of the informal sector is market-driven, leading to highly adaptable and flexible demand-driven informal waste collection forces. Generally, the volume of waste generation in rural areas is smaller than in urban areas due to the different consumption habits of inhabitants caused by a generally smaller income. Depending on the country development level, the mean rural waste generation is reported between 0.1 (countries in Asia [12], the Middle East [21] and Latin America [12]) and 0.4 kg/cap/d (rural areas in Eastern Europe [22], the Middle East [21, 23], Asia [24] and Africa [12]). The waste generation in rural areas increases rapidly up to 0.9 kg/cap/d when a touristic infrastructure is installed, becoming comparable with urban waste generation rates in developing countries, as documented for instance from Cyprus [25] and Romania [26]. In a variety of countries, only a small share of rural population has access to waste collection services [27]. Usually, informal waste collection is carried out by poor and marginalised social groups who decide for waste picking for income generation and some even for everyday survival [28].

Although urbanisation takes place in rural areas, still there are typical rural waste streams caused from rural industries like agriculture. Rural industries create waste that can be problematic to manage, like silage wrap, chemical drums and chemicals. Anyhow, those materials are not of interest for potential informal collectors as they cannot be valorised by them. As in urban areas, the main focus of waste pickers of the informal sector is on recyclable materials, especially metals and plastics, sometimes also glass as well as paper and cardboard. The waste generation rates in rural areas of developing countries are quite comparable in the range of 0.3 (Shah et al., [2, 29], for rural areas in India) up to 0.8 kg/cap*day, as reported from several sources. In countries where rurbanisation goes faster, the waste generation rate is in the upper range, for instance 0.75 kg/cap*day (with a content of mineral recyclables of about 22%) in Iran [30].

Economic activities performed by rural populations linked to informal trading and markets have not received a broad attention in the literature [31]. Thus, the question of the present investigation is the role of the informal sector in a rurbanised environment. Are there differences in the waste management activities of the informal sector in cities and urbanised rural areas?

2. Research methods

The methodology for data collection consisted of a background analysis (with a literature review), complemented with the collection of empirical evidence, field interviews and partially local field analysis. For data collection, the interviews included the following questions:

- Which are the rural waste generation rates, especially in comparison with those of urban areas?
- What is the waste composition in rural areas?
- What is the percentage collected by informal waste pickers?
- General organisation of the rural collection systems, and especially the informal sector (informal waste pickers on the streets/landfills)?
- What kind of waste do informal waste pickers collect?
- Are they an official part in the official waste management system?

Generally, the status of the informal sector is hardly documented in the literature, and most of the available data on the informal sector were collected for urban areas. Some information for rural areas is from Latin America (Colombia, Brazil), and from Africa, which was collected for this study. The reason for the poor documentation is supposed to be the informal status of the waste pickers and their 'hiding' from the statistics. To obtain information about the informal waste pickers in the rural areas, the information was collected directly by a questionnaire from the following countries (sorting in alphabetic order): Austria, the Czech Republic, Germany, Jordan, Mexico, Nepal, South Africa and Vietnam. **Figure 1** shows the location of the investigated countries in the UNICEF map of the urbanised population percentage by

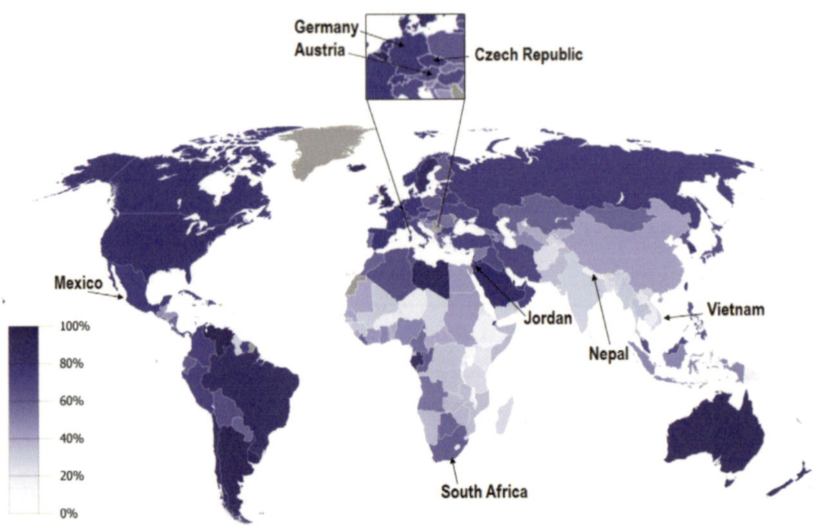

Figure 1. Urbanised population percentage by country in 2006. Map source: UNICEF, The State of the World's Children 2008 (p. 134) [32].

country in 2006 [32]. As is visible from the map, the urbanisation percentage in Austria, the Czech Republic, Germany, Jordan and Mexico was high (up to 80%), while South Africa's level reached approximately 50%, Vietnam 30% and Nepal 10%.

The information was collected from the Czech Republic, Mexico, Nepal, Vietnam, South Africa through the indicated information sources and methods:

Austria: local data collection from primary and secondary sources, as well as information collection at a Resource Management Workshop in Austria in April 2017,

Czech Republic: local data collection from primary and secondary sources,

Germany: local data collection from primary and secondary sources, as well as interviews with representatives of local waste management authorities,

Jordan: local data collection from primary and secondary sources, as well as information collection in March 2017,

Mexico: local data collection from primary and secondary sources,

Nepal: local data collection from primary and secondary sources, information received from the Solid Waste Management and Resource Mobilization Centre (SWMRMC) in Nepal. The SWMRMC made an investigation in each municipality [33], categorised them into urban and rural wards as smallest administrative unit. The rural wards are characterised through lesser population density than urban areas and without commercial activities, where the representative households in each municipality were selected randomly by employing the right-hand-rule technique (Asian Development Bank [48, 49]).

South Africa: local data collection from primary and secondary sources, as well as field research and interviews with the waste pickers from the informal sector in February 2017,

Vietnam: local data collection from primary and secondary sources, as well as information collection at the National Farmers Union in January 2017.

By nature, the collected data had inhomogeneous composition, due to two reasons: firstly, the data availability strongly varied in the countries, and even concerned the type of data; secondly, not all types of data could be collected from all countries. Anyhow, for the recent scope of the investigation, the data were sufficient, as the aim of the chapter is to give an overview on the variety of settings for the informal sector.

3. Investigation results

The results are a summary of the collected data for each country, which gives information on the collection scheme, as well as the involvement and the activity of the informal sector in the respective countries. Generally, it was observed that the informal sector existed in urban and rural areas; even the quantity of waste collected was smaller in the rural areas. Furthermore, settlements of the informal sector can be found in the areas of communal dumpsites and landfills, collecting already recyclable materials before the waste goes to the dumpsite and landfill. Furthermore, the extent of the activity of the informal sector depends on the type and structure of the collection system in the country. Usually, when the collection system is not a selective system separating the recyclables, there is a larger activity of the informal sector. This is usually the case in low- and middle-income countries. In high-income countries like Germany in Central Europe, a real informal sector does not exist.

3.1. Situation in high-income countries: Germany, Austria, and the Czech Republic

According to OECD information, Germany has 80.6 million inhabitants with an average household income of 34,700 US$. The average waste generation rate varies between 0.65 and 1.37 kg/cap/d in rural areas, in comparison to urban areas with waste generation rates between 1.37 and 2.2 kg/cap/d, having a total average of 1.68 kg/cap/d [34]. The waste collection system is a selective system, which separates recyclables (glass, paper, plastic (PET) bottles, other plastics, metals and biodegradable waste) from residual waste. The waste management system is operated by municipal or communal operators, and only exceptionally by private operators, a situation which applies for rural and urban areas.

The activities that can be considered as a type of informal sector activity are some private poor people who collect bottles and cans from the streets in order to transfer them to the bottle deposit refund system, which exists in an automated way in each supermarket or rural discounter. For instance, in Germany, the refund for one PET bottle or one metal can is 0.25 € (0.27 US$), while the refund for a glass bottle is 0.08 € (0.09 US$). The deposit refund was calculated according to the environmental risk (PET bottles) or the material value (metal cans), and is equal in rural and in urban areas. In Germany, a deposit-refunding system for bottles of alcoholic beverages does not yet exist, but it is under governmental preparation (status as of February 2017). For that reason, the waste bottles most often found in the environment are bottles of alcoholic beverages which are not of interest to informal collectors. Formal collectors provide glass containers, where consumers put those types of bottles and packaging glass for material valorisation. The existing system fulfils the scope of a clean environment, and private bottle pickers are the exception. Furthermore, all municipal landfills were closed by law in 2005, and landfills for the disposal of untreated municipal waste have not existed anymore since then. All generated waste has to undergo a pre-treatment, before recycling or re-using as priority options, and only hazardous waste is disposed of. The described situation is the reason that informal waste collectors on landfills do not exist at all in Germany.

Other waste is not collected by private waste pickers, as all waste streams are collected in selective collection schemes through formal collection systems of the municipalities, which is valid for urban as well as for rural areas, centralised in civic amenity areas. In zones which are close (up to 50 km) to the East European border (with Poland or the Czech Republic), there are informal East European waste collectors (especially from Romania, Hungary, Poland and the Czech Republic) [35] waiting outside the civic amenity centres to collect usable waste directly from the customers who are bringing waste to the centres. Usually, they are collecting household appliances, textiles, toys and other items for children, sports equipment, electrical appliances such as TV sets, washing machines or refrigerators, tires, scrap metals and other bulky waste, for instance from furniture. The transfer of this kind of waste is free of charge, and even it is not really allowed, it is tolerated, and in this way informal by nature.

A comparable situation does exist in Austria. The country has 8.5 million inhabitants with an average household income of 45,500 US$. The average waste generation rate being 1.58 kg/cap/d is slightly lower than in Germany, having generally comparable dimensions to Germany for rural and urban areas. A visit to Austria in April 2017 indicated a certain percentage of waste bottles in the urban area spread around public collection bins while there was nearly no waste in rural areas in the environment. The result of the interview indicated that there is a comparable deposit refund system like in Germany, but obviously it appeared not to be efficient everywhere, maybe because the deposit refund was too small. In Austria, bottles are partly pledged. For simple reusable beer bottles, 0.09 € (0.10 US$) are refunded and 0.36 € (0.40 US$) for special types of beer bottles. For reusable PET bottles as used by some mineral water and lemonade manufacturers, a 0.29 € (0.33 US$) deposit is charged, as well as for 1-l mineral water glass bottles. Anyhow, relevant informal activities are practiced for the same materials as in Germany, which can be considered as a particularity of a high-income country. Also an informal waste transfer from Austria to Eastern Europe countries by informal waste collectors does exist. According to Obersteiner et al. [35], 69% of the informal waste collectors in Austria originate from Hungary and 19% from Austria. The rest comes from Bulgaria, the Czech Republic, Slovenia, Slovakia and Romania. Istvan et al. [36] reported that informal waste collectors from Hungary even travel for waste collection to the Netherlands. According to Obersteiner et al. [37], a verification at the Hungarian border showed that the collected items were 47.21% by volume of furniture, 18.77% by volume of electrical appliances and 13.19 Vol% of metals.

Also, the Czech Republic is considered a high-income country; even waste collectors from the Czech Republic come to the neighbouring countries like Germany and Austria, as the income there is even higher. The Czech Republic has 10.5 million inhabitants and an average household income of 17,542 US$. According to the income, which is proportionally lower than in Germany or Austria, the average waste generation rate is also lower: 0.8 kg/cap/d, and also significantly lower than the EU average of 1.3 kg/cap/d. No refund is applied for aluminium cans or plastic bottles in the Czech Republic, only some kinds of glass bottles are refunded for 3 Kč (approximately 0.11 US$). That is why the 'secondary' collection of this type of waste is negligible there.

The Waste Law of the Czech Republic orders the municipalities and communes to arrange waste collection places so that some parts of the waste (esp. glass, paper, plastic, metals and biowaste) should be collected separately. All rural areas are administrated by their central municipalities, meaning that law and waste management in rural and in urban areas are the same. The informal waste collectors are active in the Czech Republic, even being gypsies like in Romania and Hungary. The waste proportion collected by them is finally included into the waste that is recycled by the recycling companies and in that way included into the statistics. The informal sector usually collects metals that can be simply sold. They sometimes also steal some metal parts of working systems (electrical wires, railway security systems, monuments, sewer covers, etc.) and sell them as metal waste. They are not foreseen to be a part of the official system even there are some laws and procedures to prevent them. Generally, the informal waste pickers are much more active in the poor areas of the country (Northern Bohemia or Northern Moravia) than in the rich regions.

An investigation carried out by Tydlitatova et al. [38] in several rural communes in the Czech Republic on the impact of the implementation of the system pay as you throw (PAYT) showed that the villages, which applied Local Tax system, produced 47% more of mixed municipal waste. The villages that applied Local Tax generated an average of 0.52 t of mixed municipal waste per 5 years, and more than the villages that applied the fee by Act on Waste [38]. The results according to Tydlitatova et al. [38] are given in **Table 1**. The example from the Czech Republic shows that not only the average household income has an impact on the waste generation rate but also the system of payment of waste fees. Higher fees have a regulating impact and cause lower waste generation rates.

Municipality	Population (2011)	Applied waste law (2011)	Fee per person or dustbin	Fee per person or dustbin	Distance to landfill, km
Horažďovice	5578	Local Tax	CZK 600/person	US$ 24/person	43
Horoměřice	3335	Local Tax	CZK 480/person	US 19/person	6
Jílové u Prahy	4222	Local Tax	CZK 500/person	US$ 20/person	1
Mnichovice	3069	Fee by Act on Waste	CZK 1750/120 l	US$ 70/120 l	35
Psáry	3331	Fee by Act on Waste	CZK 2145/120 l	US$ 86/120 l	53
Říčany	13,499	Contractual form by Act on Waste	CZK 2520/120 l	US$ 101/120 l	36
Statenice	1261	Local Tax	CZK 600/person	US$ 24/person	6

Table 1. Waste management system in several rural municipalities [38].

In the following discussion, the focus is on low- and middle-income countries which all face the issue of informal waste pickers, in the urban as well as in the rural areas. The respective countries are considered in alphabetic order.

3.2. Jordan (lower middle-income country)

Jordan is a lower middle-income country in the Middle East, with an original number of inhabitants of 6.5 million in 2013 (data of OECD), which increased through migrants from Iraq and Syria recently by at least 2 million; 21.2% of the inhabitants live in the rural areas [39]. The annual average household income is approximately 5160 US$, with a large variation. The average waste generation is 0.9 kg/cap/d in urban areas and 0.6 kg/cap/d in rural areas. Jordan is quite densely populated, and the existing informal waste collection sector has undergone an even higher competition after a large number of migrants entered the country to search for possibilities to ensure their income for living, as reported in interviews in March 2017. This kind of situation was recently also observed in Turkey, where the existing

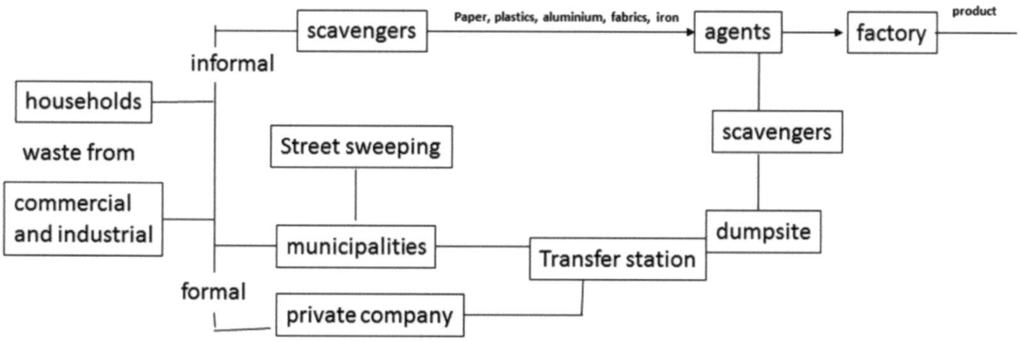

Figure 2. Flow chart for solid waste streams and scavengers role in Jordan ([39], adapted).

well-organised informal sector got quite under economic pressure caused by a stronger competition. Resource recovery and recycling are practised in a limited way, even those of urban areas are clean and free from street waste. In the rural areas, there is a higher percentage of waste beside the roads, and it is obvious that there is cleaned or collected much more seldom.

A well-documented study on the informal sector in the rural and rurbanised environment of Jordan was provided by Aljaradin et al. [39], which analysed the informal recycling activities carried out by a scavenger in the Tafila region of Jordan. The general situation is given in **Figure 2**, and it is a typical situation for a variety of low- and lower-middle-income countries.

There is no legislation which forbids scavengers to pick and recycle waste but the Ministry of Social Development always tracks them for children working as waste pickers [39]. The informal recycling in Jordan was estimated to be around 10% from the total municipal solid waste (MSW) generated. As shown in **Figure 2**, their activities are carried out before the solid waste reaches the final disposal sites for the separation of recyclable materials, but the majority of informal collection is done at the disposal sites. The informal waste collectors are welcomed as they reduce the cost of formal waste management systems. The materials most often collected are aluminium, plastic, paper, cardboard, glass, copper and iron [39]. The average quantity collected by 100 scavengers per day is reported with 150-kg soft drink cans, 5-kg aluminium stripes, 2-kg copper wires and 90-kg scrap metals.

The average waste composition contains biodegradable waste (52%), plastics (17%), paper/cardboard (14%), glass (3%), metals (1%) and others (17%) (Karak et al. [40]). The composition of the scavenger crowd in the Tafila region is 99% men and 1% woman [39], with 80% being less than 25 years old. The majority of the informal waste pickers in Tafila (78%) obtain a monthly income of >250 € (268 US$), the others <250 €. As Aljaradin et al. [39] reported, scavengers usually have no concept of the essential role of their work in the waste management activities, and their social status is very low.

3.3. Mexico (upper middle-income country)

Mexico is a country in Latin America with 122 million inhabitants. The annual average household income is 12,800 US$. The average waste generation in rural communities is 0.68–1.09 kg/cap/d [41]. In other studies, carried out in rural communities in Mexico the interval found is between

0.28 and 0.58 kg/cap/d [41, 42], indicating urbanised behaviour. It can be assumed that the differences result from the consideration of agricultural wastes. A study carried out in eight communities from Michoacan, Mexico [42], points a composition of 44% of food scraps, 8% of yard trimmings, 2% of cardboard, 2.8% of paper and 0.6% of textiles [41]. In comparison, the per-capita MSW generation in the urban area ranged from 327 to 361.35 kg/inhabitant/year from 1995 to 2012 [43].

As in other developing countries, also in Mexico, the informal sector exists, which is concerned with the recovery of waste, but an investigation to quantify the contribution regarding the recovery of recyclables [44] would be necessary. According to Taboada-González et al. [41], in some rural communities of Mexico, waste collection is provided by the municipality through the Department of Waste Management (DWM) at no charge. The waste is collected once a week at the curbside where residents place their garbage bins. Afterwards, waste is disposed of in each community's dumpsite. The percentage of coverage of waste collection services in the rural area is 60%, making it clear that the DWM does not totally collect the waste generated by the communities, being inefficient in most of the cases. The rest of the waste is usually mismanaged and burned outdoors or discarded at ravines, uncultivated land and canals. Also, an unquantified fraction is collected by informal collection services that offer their services in exchange of a gratuity. Also in Mexico a deposit refund system exists.

Figure 3. Informal refuse collection in Netzahualcoyotl, Mexico. Photo by Medina [45].

In Mexico, scavenging and informal refuse collection (IRCs) is very common (**Figure 3**) [45]. In many cases, rag pickers recover some valuable materials (aluminium, tin can and ferrous waste) and the rest is dispersed to be burned outdoors. Waste picking is done near the source, that is, after collection has taken place at the generating sources and previous to being transported to the dump or landfill. The most common way of selling the collected material is directly to the companies that attend the site daily [43]. Materials such as aluminium, tin cans and ferrous waste are collected by waste pickers in rural communities [41]. Waste pickers working in the landfill collect mainly plastic (PET), also aluminium cans, plastic (HDPE) and metals. The material is selected for collection in terms of the market for each product [44]. According to Medina [43], in some towns, informal refuse collectors pick up garbage and charge each home a fee between US$ 0.10 and 0.50 (**Figure 4**). In many cases when they operate in a place far from the municipal disposal sites, they take the collected waste to privately operated transfer stations and pay a fee of US$ 1–4 for unloading wastes there, depending on the amount. Hence, in addition to collection fees, they recover recyclables from the wastes, which, considering the fees they pay, results in an average income of US$ 9–15 a day, which is between three and five times the minimum wage. Being so, in many cases IRC is a highly paid activity.

Figure 4. Informal refuse collection in Tultepec, Mexico. Photo by Medina [45].

	MXN Peso/kg	US$/kg
Paper	3.00	0.156
Newspaper	1.50	0.078
Glass	0.60	0.031
Plastic PET	3.00	0.156
Cardboard	2.00	0.104
Aluminium	17.00	0.884
Food tins	17.00	0.884
Metal	2.00	0.104
Magazines	3.00	0.156

Table 2. Recyclables prices in Mexico [46].

In many towns not just in Mexico but in Latin America, the informal sector has been used as a semi-official tool to bring services to low-income areas, which offers a more open system, responding to basic needs and demands [44]. **Table 2** shows the recyclable prices in Mexico [46].

Generally, there is a direct relationship between producers and consumers in the informal sector, requiring low capital, which allows for more rapid growth. However, this peculiar nature of the informal sector makes monitoring and regulation more difficult, for which it has resulted in the inefficiencies previously mentioned. Hence, as stated by Medina [45], incorporating informal collection services into the municipal waste systems and formal programmes could bring some control over their operations and stop illegal dumping.

3.4. South Africa (upper middle-income country)

The country located in the Southern Africa has 70 million inhabitants with an annual average income of 5845 US$. The average waste generation is 1.7 kg/cap/d in urban regions and 0.35 kg/cap/d in rural regions [47]. Results of a topic-related research of the Council for Scientific and Industrial Research South Africa on the informal waste sector indicated that between 60,000 and 90,000 waste pickers earn a livelihood from the recovery of recyclables from municipal waste in South Africa. This intensive informal sector, which especially also works in the rural areas, provides a valuable, and low-cost recycling solution. While the informal sector in the urban areas is going to be formalised step by step, the informal sector in rural areas is mainly living from the activity of private recycling companies. Generally, the situation in the urban area is easier and more economic for an informal waste picker than in the rural area. In the urban area, for instance, in Bloemfontein, the informal sector was somehow formalised through green T-shirts, which must be bought, and represent the official allowance to collect recyclables. In this way formalised, the waste picker can act as glass recycler and earn up to 12,000 ZAR/y (approximately 923 US$/y).

In the rural areas, the informal collection is a very difficult job. Usually, the informal sector collects the recyclables at landfills, means on landfills with an informal allowance to enter them, or in front of the landfill at the entrance, or on the rural road, which is connecting the landfill. There, the informal recyclers even stop cars, which are on the way to the landfill. Besides those activities, also conventional collection activities in the villages do exist, even they are not the majority of the activities leading to income for the informal waste recyclers. Generating income with informal activities in South Africa is a quite unpredictable activity. As the waste collectors reported in interviews in February 2017, they do not know when the private waste recycling company sends the trucks to collect the waste of the informal sector, which usually happens twice a year, sometimes only once a year, but the date is not announced.

This leads to the situation that the informal waste collectors need to establish waste storage sites (usually outside landfills), where the recyclable waste fractions are already pre-sorted and packed to be ready for the collection by the recycler in each moment. Such constellation leads to informal settlements for the purpose of waste collection and manual pre-sorting. The payment is small: 2 ZAR per kilogram metal (0.15 US$), 1 ZAR per kilogram plastics (0.08 US$) and 1 ZAR per kilogram glass (0.08 US$). If the collection activity goes properly, and the private recycling company sends the collection truck, an informal waste recycler can earn approximately

6000 ZAR/y (approximately 461 US$/year). This annual income of an informal waste collector in the rural areas compares to half of an average monthly income of a worker in an urban area. There are nearly no women doing this kind of job in the rural recycling settlements.

Figures 5–7 show impressions of an informal waste recycling settlement in the rural areas of eMalahleni, taken in February 2017.

Figure 5. Informal waste-recycling settlement in the rural areas of South Africa, close to eMalahleni: PET collection (photos taken by the authors on 26 February 2017).

Figure 6. Informal waste-recycling settlement in the rural areas of South Africa, close to eMalahleni: glass collection (photos taken by the authors on 26 February 2017).

Figure 7. Informal waste-recycling settlement in the rural areas of South Africa, close to eMalahleni: living conditions at the informal recycling village (photos taken by the authors on 26 February 2017).

3.5. Nepal (low-income country)

Nepal is a country in Eastern Asia which consists of mountains, hills and a lowland region which is called Terai. The country has 28 million inhabitants and an annual average household income of 701 US$. For the study of the Solid Waste Management and Resource Mobilization Centre [33], a total sample size of 3330 households from 60 municipalities in the rural areas selected from all ecological zones was considered, having 55 households that gave an average per-capita household waste generation of 0.12 kg/cap/d [33]. The data base for Nepal shows that the household waste generation rates in new municipalities varied depending upon the economic status. The average waste generation correlates with the monthly available household income. Households with a monthly budget of NRs ≥40,000 (about 389 US$) generate 0.88 kg/day, in comparison to 0.4 kg/day for households with a monthly budget of less than NRs ≤5000 (about 49 US$) [48]. The results of the study indicated a per-capita household waste generation from a minimum of 0.07 kg/cap/day (Bheriganga Municipality) to a maximum of 0.22 kg/cap/day (Bhojpur Municipality) [48].

The characteristics of MSW collected from any area depend on various factors such as consumer patterns, food habits, cultural traditions of inhabitants, lifestyles, climate, economic status, and so on. Composition of urban waste is changing with increasing use of packaging material and plastics. The average household waste composition investigated in 60 municipalities in terms of the eight determining waste components (organics, plastics, paper and paper products, glass, metal, rubber and leather, textiles and others like inert and dust) is presented in **Figure 8** [33].

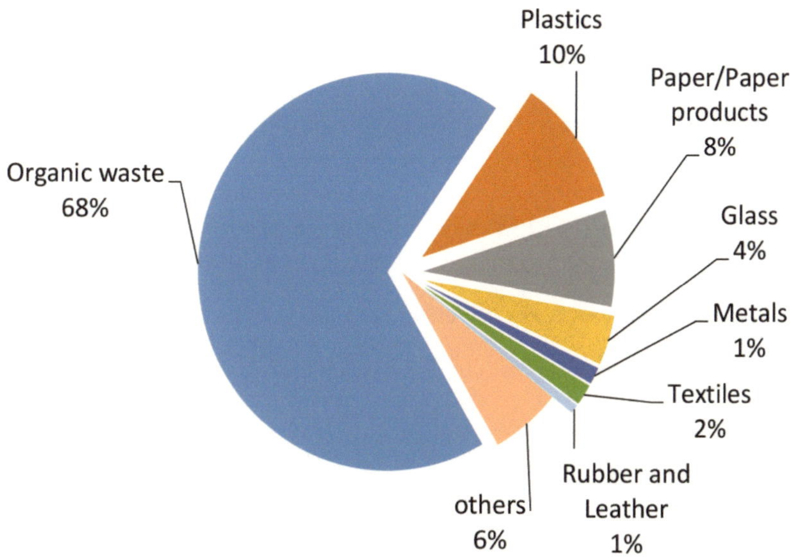

Figure 8. Average composition of household waste of 60 rural municipalities, with average values by % wet weight [33].

The average composition of household waste comprises as highest fractions organic matter (68%), followed by plastics (10%), paper and paper products with 8% as well as other types of waste with 8.6% [33]. The rest, being below 4% wet weight, was glass, metal, rubber and leather, and textile components [33].

Of total surveyed households from 60 municipalities, 51% responded that they are practicing segregation of waste at sources, which is higher than that of survey findings from 58 municipalities conducted in 2012 [33]. The higher segregation at sources in new municipalities is because of rural nature of these municipalities where almost all households were found to segregate kitchen waste for their own purpose, for example, feeding cattle and using for traditional type composting, and so on. Moreover, only 33% surveyed households have composting practices of segregated waste while 37% do not have such composting practices in their households, which means that the segregated waste at source is mixed again during collection and transportation due to the lack of separate collection and treatment methods [48].

In many of the new municipalities, a solid waste collection system does not exist, and if the system exists, it is not satisfactory due to unscientific composting, or open burning, or throwing the waste in the open space around [33]. Only 2% of surveyed households sell segregated non-biodegradable fraction to informal sectors. 52% respondents told that they do burn of segregated non-biodegradable waste like plastics and papers, while remaining either throw into road drains or do both [49]. Collection, city cleaning and sweeping do not happen on a daily basis [33], and only main market and roads are served daily. Other areas are served intermittently, from twice a week to twice a month [33]. Many areas in the rural environment are neglected due to inefficiency and inadequacy of service [49]. Although a concept of material recovery from MSW with legal provisions of sorting waste at sources has been already introduced in Nepal through the new solid waste management act (SWMA) promulgated in July 2011, no formal municipal waste recovery and recycling programme exist in the municipalities [48]. Because of municipal budget constraints, municipalities try to create a sound budget without increasing cost-efficiency option, but arrive at the point that MSW has become environmental, financial and social burden to each municipality [33]. The conclusion of the investigation was that only 2% of surveyed households sell segregated reusable and recyclable fraction to informal sectors, being considered as very minimum resource recovery activities in the surveyed municipalities [33].

Figure 9 shows informal workers collecting recyclable materials, Belabari Municipality, Nepal.

Figure 9. Informal workers collecting recyclable materials, Belabari Municipality, Nepal (photo source: Solid Waste Management Technical Support Center [33], with friendly permission of SWMRMC Nepal).

3.6. Vietnam (low-income country)

Vietnam is a country in South East Asia, having a population heading towards 90 million inhabitants. The annual average household income was 1912 US$ in 2014. The economy of Vietnam is agriculture with paddy rice as major crop cultivated on 4.5 million ha land. In addition to the main product, rice grain, by-products such as rice straw and husk (renewable resource) are also produced, estimated to be around 38 million tonnes of straw and 6 to 7 million tonnes of husk per year for whole Vietnam [50].

The country is very densely populated. People live in all types of organisational forms, covering sizes from the rural community to megacities. The agricultural sector in the rural areas plays a fundamental role in Vietnam, as it provides the nutrition for the growing population. The average waste generation in the rural area is 0.4–0.5 kg/cap/d (excluding agricultural wastes). The organic content of the generated waste is very high, up to 90%. Plastic bags form a percentage up to 15%. A serious problem is the percentage of hazardous waste, especially bags, bottles and cans which contained pesticides. In private households, the waste is often burned. Only a small percentage is composted.

The waste collection is organised in different ways in the provinces, usually comprising two levels:

(a) Waste collection from the households through companies, NGOs (e.g. Farmers Union, Women Union, Veterans) or private waste pickers (informal sector). The waste collection fee is stipulated by themselves, as well as the determination of the collected fees (payment for employees or investments). Informal waste pickers exist everywhere in the rural areas, and they mainly collect plastics, paper and metal. Only a small percentage of the collected waste is recycled, the majority is put on landfills. In areas, which are far from the official collection and recycling infrastructure, the waste is deposited into illegal dumpsites.

(b) Waste collection at the landfills of the province. The central waste management company URENCO allows private waste pickers to collect recyclables from the deposited waste and pays for this service.

Vietnam produces many biodegradable wastes in the rural areas, which are mainly not recycled in the current state. The biomass contributes to the rural waste generation, and even it is recyclable, it is not yet properly valorised. For the informal sector, it is not an interesting waste stream. Currently, in the urban and rural areas in Vietnam, a bottle deposit refund system for beer and soft drink bottles does exist. The bottles refund for one box with 20 beer or 24 soft drink bottles is 20.000–40.000 VND (0.88–1.76 US$). The recyclable materials are bought by collectors from households with the following prices:

Cardboard: 3000–4000 VND/kg (0.1–0.18 US$)

Paper: 4000–5000 VND/kg (0.18–0.22 US$)

White cleaned covers from nylon and plastics: 12,000 VND/kg (0.53 US$)

Coloured cleaned covers from nylon and plastics: 10,000 VND/kg (0.44 US$)

Dirty covers from nylon and plastics: 2000 VND/kg (0.09 US$)

PET bottles: 4500 VND/kg (0.20 US$)

Iron scrap: 4800 -10,000 VND/kg (0.21–0.44 US$)

Aluminium scrap: 20,000 VND/kg (0.88 US$)

Copper scrap: 60,000–90,000 VND/kg (2.65–3.97 US$).

Figures 10 and **11** show informal workers collecting recyclable materials in the Mekong region in Vietnam.

Figure 10. Informal workers collecting recyclable materials, Mekong region, Vietnam (photos taken by the authors on 14th august 2017).

Figure 11. Informal workers collecting paper and cardboard, Mekong region, Vietnam (photos taken by the authors on 14th august 2017).

4. Discussion

The informal collection of waste is a phenomenon that results from social differences within society and the population. Therefore, it is not surprising that the perception of the activities of informal waste collectors in the scientific literature refers to developing countries and emerging countries, since social differences are more pronounced. These informal waste management systems in low- and middle-income countries usually exist in parallel with formal waste management systems, and this applies for urban as well as rural areas and might be considered as a result of rurbanisation. The case studies show the development of the informal sector as an important part of the waste management activities, when a country starts to develop. With increasing economic development, the importance of the informal sector is shrinking step by step in relation with the improvement of the formal activities. Even this development goes faster in urban areas; the conclusion applies also to rural areas. Although organic waste is the main waste stream in rural areas, there is a relevant proportion of informal activities on recyclables like metals, plastics, papers and glass.

One of the main focuses of the formal waste management activities in the urban areas is to find solutions for the inclusion of the informal sector into the formal activities, and in this way, it is formalisation. In order to support the consideration of formalisation options of the informal sector, a further literature search was carried out. The aim was to reconcile the approaches used in other countries and to consider as far as possible a wide range of ideas. Thus, in principle, the following approaches exist in the literature:

- Incentive systems for the disposal of certain wastes.
- Umbrella organisation for informal collectors outside the waste regime.
- Incorporating informal waste collectors into a commercial waste management company.
- Establishment of an umbrella organisation within the waste regime.
- Second-hand goods trading through municipal waste management.
- Consideration in national ReUse concepts.

The following basic options for the integration of the informal sector were determined in the literature research in the form of case studies (in accordance with the investigations of Hold, [14]):

(A) Involvement of informal waste collectors with the municipality by means of a contracted subcontract to an unionised unit, for example, for certain geographically defined administrative units and/or for the lower-income areas. An example is Maputo (Mozambique): The informal waste collectors supply the recyclable waste to a centre for the sorting and pre-treatment of valuable substances. The recycling centre sells high-quality, recycled plastics to the local recycling industry and is independent of external financial support.

(B) Involvement of informal waste collectors by specialising in a specific type of waste. In Delhi and Bangalore (India), numerous informal workers recycle electrical and electronic waste. The informal recyclers received training in the risk of their work and appropriate recycling techniques. Finally, several small recyclers of electronic waste joined together and were able to register themselves officially. In the treatment of hazardous substances, the new company cooperates with an experienced, formal recycler. The newly founded and registered company has established itself on the market and offered its employees an improved job situation.

(C) Specified waste refunds for certain recyclable waste types and quantities. In Bogota (Colombia), informal waste collectors receive 28.78 US$/tonne of waste collected at officially authorised collection points in the city; 13,754 informal waste collectors are found in Bogota, 58% of them are women; 1200 tonnes of waste are collected daily by them. The average income per collector is 3.41 US$/d.

(D) Official recognition of the waste collector is a profession. In Brazil, the waste collector was recognised as a profession. There are three types of organisations: (1) unorganised and anonymous waste collectors, which are not affiliated with any organisation, (2) organised waste collectors organised in cooperatives and associations, and, as a rule, at least 10 years of professional experience, as well as contract-bound waste collectors, mainly working in scrap metal, in metal works and also in the municipal sector. In Brazil, there are 229,568 waste collectors, of which 67% are men, 25% are between 50 and 64 years old and 7% are over 65 years old. Only 14% of them have a school-leaving certificate. Approximately 4.5% work in a formal contract, which include 11,781 people. The contractually employed persons have a median income of about three to four times higher.

(E) Another possibility to improve the problematic situation of informal waste management would be to take account of informal collectors with the introduction of a national re-use system. The informal collectors could fulfil the need for new capacity in waste collection centres, such as storage areas and labour, or take over the transports between the municipal collection points and the socio-economic enterprises, or assist the employees of social economy enterprises with their repair knowledge.

(F) Integration of the informal sector into the development of re-use and repair networks in co-operation with socio-economic integration companies. In this formalisation, the informal collectors function according to Scherhaufer et al. [51] as transporters for socio-economic enterprises. A separate collection of reusable items is carried out in the waste collection centres in question, which are then transported to the ReUse plants by the informal collectors. The delivered items are then sorted according to their functions in ReUse operation. They are divided into items with or without the need for repairs [14].

The mentioned general options could be applied also in the rural areas; even the probability of the feasibility of some particular options is higher (in accordance with the investigations of Hold, [14]), as there are:

- Option (A) could be applied also on communal basis in rural areas.
- Option (B) appears not as an applicable option as the waste volumes in rural areas are smaller, and with this the proportion of usable recyclables is smaller. The example from South Africa shows a remarkable volume of recyclables needs time to be collected in the rural areas.

- Option (C) is an interesting option, which could allow for a constant cash flow or income.
- Option (D) is a challenge. It would be useful to recognise waste recycling as profession, especially for the informal sector. This option could be combined with the special clothing of the waste recyclers in order to recognise them faster as officially working people.
- Option (E): this option would work also in rural areas, but it requires as precondition that the country is already in a certain developed stage.
- Option (F): this option applies to a higher-developed society which already has a developed waste management infrastructure.

5. Conclusions

Scope of the current investigation was to collect data from literature and through field studies in order to obtain information on the informal sector activities in the rural areas, working in a rurbanised environment. A general conclusion from the questionnaires and field visits is that the informal sector exists also in rural areas; even the generation of recyclable waste is smaller than in urban areas. Therefore, the income of rural informal waste pickers is lower than that of urban waste pickers. As the informal sector in the rural area is usually concentrated near the landfills, they use recyclable materials going to the landfill in several ways to make their living. Usually they collect metals, glass, PET bottles and sometimes also papers. Potential differences in the waste management activities of the informal sector in cities and in an urbanised rural environment can be stated at this investigation stage that the urban sector shall be usually formalised at a certain development stage, while this is usually not yet the case in the rural area. Further, like in other commercial sectors, the income in rural areas is usually less than in urban areas. The percentage of women in this sector is negligibly low.

Most of the middle- and low-income countries deal with an informal waste sector. And usually, each respective country faces a number of unique socio-economic and political circumstances that may influence the integration of the informal sector into a formal secondary resources economy. Anyhow, one question in this regard is: What model of social inclusion of waste pickers would be most appropriate in the respective country, means integration and/or formalisation? A discussion in a recent workshop at the Chamber of Commerce, held in Istanbul in October 2016 was on the subject of the inclusion of the informal sector into the official waste management system. The informal sector also participated. Surprisingly, not all of the members of the informal sector agreed to be included in the formal waste management organisation. Most of them told the freedom of their working conditions as reason.

Having in view the process of rurbanisation and the economic development of the low- and middle-income countries, informal waste pickers are at present an important part of the system. It is to be expected that with increasing economic infrastructural development, their relevance will be decreasing on the long term, but not necessarily. In Vietnam, for instance, the informal sector is included in the waste management system as official power already to collect recyclables from landfill (e.g. Nam Son in Hanoi). On the other hand, it might be possible that the informal sector covers especially rural regions, which are less developed in infrastructure and/or which are far from the next recycling centres and not economically manageable with formal waste management activities. In such case, the informal sector could be able to manage those regions with its technical means, for instance horse carriage (like in several East European countries), or smaller motorised vehicles. Anyhow, the main waste stream in rural areas which will not be managed by the informal sector is the organic waste.

Acknowledgements

Collection of data on the informal sector is not feasible without the support of many people, starting from the waste pickers through recycling companies to waste management authorities. We are grateful to all of them. Thanks to Pieter, Abraham, Vusi, Skumbulu, Sam, Given and Mandla from an informal collection location in South Africa; thanks to Prof. Dr. Christian Wolkersdorfer who gave technical support to conduct the interviews with the informal waste collectors in South Africa. Thanks also to Dr. Sumitra Amatya from the Solid Waste Management and Resource Mobilization Centre in Nepal, to Mrs. Birgit Dietz from the Vogtlandkreis Waste Authority in Germany; thanks to Ing. Danuše Hráská from The Czech Environmental Inspectorate. We hope that with this contribution we were able to put the situation of the informal sector in the awareness focus of the public and the special situation which exists in rural areas.

Author details

Petra Schneider[1]*, Le Hung Anh[2], Jan Sembera[3] and Rodolfo Silva[4]

*Address all correspondence to: petra.schneider@hs-magdeburg.de

1 Department of Water, Environment, Civil Engineering, and Safety, University of Applied Sciences Magdeburg-Stendal, Magdeburg, Germany

2 Industrial University of Ho Chi Minh City, Ho Chi Minh City, Vietnam

3 Faculty for Mechatronics, Informatics and Interdisciplinary Studies, Technical University of Liberec, Liberec, Czech Republic

4 FGlez Consulting & Associates, Mexico City, Mexico

References

[1] Sorokin P, Zimmerman CC. Principles of Rural-Urban Sociology. Vol. 652. New York, NY: Henry Holt and Company; 1929. pp. xv

[2] Parsons T. The Structure of Social Action: A study in Social Theory with Special Reference to a Group of Recent European Writers. New York, USA: Free Press; 1949

[3] Chapuis R, Brossard T. The demographic evolution of rural France (1968-1982). Journal of Rural Studies. 1989;**5**(4):357-365

[4] Banon M. The rural environment. Economie et Finances Agricoles. 1980;**1980**:63-68

[5] Barbichon G. Industry and rurbanization: Sociological aspects. Economie Rurale. 1977;**117**:28-34

[6] Bauer G. Suburban countryside, spreading towns: How much is known of the "rurban" phenomenon? Economie Rurale. 1977;**117**:13-16

[7] Borras SM, Jr. Special issue: Critical perspectives in agrarian change and peasant studies. Journal of Peasant Studies. 2009;**36**(1):5-265

[8] Boyer JC. Second homes and "rurbanization" in the Paris region. Tijdschrift voor Economische en Sociale Geografie. 1980;**71**(2):78-87

[9] Paveliuc-Olariu C. The analysis of the effects of rurbanization on rural communities in the North-East Development of Romania. AAB Bioflux. 2010;**2**(1):41-48. DOI: http://www.aab.bioflux.com.ro

[10] Kayser B. The rural renaissance. In: Colin A, editor. Sociology of Campaigns in the Western World. Paris, France: Armand Colin publishing house; 1990. p. 68. (In French)

[11] Mahajan S. Rurbanization. 2010 [Online]. Available from: http://sdmahajan.tripod.com/rurbanization.htm

[12] Hoornweg D, Bhada-Tata P. What A Waste: A Global Review of Solid Waste Management, Urban Development & Local Government Unit. Washington, USA: World Bank; 2012

[13] Chi X, Streicher-Porte M, Wang M, Reuter M. Informal electronic waste recycling: A sector review with special focus on China. Waste Management. 2011;**31**:731-742

[14] Hold F. Informelle Abfallwirtschaft in Österreich - Chancen, Risiken und Praxis, Universität Graz, Institut für Systemwissenschaften, Innovations- und Nachhaltigkeitsforschung Graz, October 2012

[15] Wilson D, Whiteman A, Tormin A. Strategic Planning Guide for Municipal Solid Waste Management. Washington, DC: World Bank; 2001 [Online]. Available from: http://www.worldbank.org/urban/solid_wm/erm/start_up.pdf

[16] Benson E et al. Informal and Green? The Forgotten Voice in the Transition to a Green Economy. London: International Institute for Environment and Development (IIED); 2014

[17] Scheinberg A, Wilson DC, Rodic L. Solid Waste Management in the World's Cities. London: UN-Habitat by Earthscan; 2010

[18] Haan HC, Coad A, Lardinois I. Municipal waste management: Involving micro-and-small enterprises. Guidelines for Municipal Managers. Turin, Italy: International Training Centre of the ILO, SKAT, WASTE. 1998. Available from: http://www.skat-foundation.org/publications/waste.htm

[19] Scheinberg A, Simpson M, Gupt Y, et al. Economic Aspects of the Informal Sector in Solid Waste Management. Eschborn, Germany: GTZ and CWG; 2010

[20] Deutsche Gesellschaft für Technische Zusammenarbeit (GTZ) GmbH. The Waste Experts: Enabling Conditions for Informal Sector Integration in Solid Waste Management Lessons learned from Brazil, Egypt and India. Eschborn. March 2010

[21] Darban Astane AR, Hajilo M. Factors affecting the rural domestic waste generation. Global Journal of Environmental Science and Management. 2017;**3**(4). DOI: 10.22034/gjesm.2017.03.04.00 Available from: http://www.gjesm.net/article_25252.html

[22] Mihai FC, Oiste AM, Chelaru DA. Rural waste generation: A geographical survey at local scale. In: Conference Proceedings of the 14th International Multidisciplinary Scientific GeoConference; Albena, Vol. 1; 2014. pp. 585-593. DOI: http://dx.doi.org/10.5593/SGEM2014/B51/S20.080

[23] Abduli MA, Samieifard R, Jalili Ghazi Zade M. Rural solid waste management. International Journal of Environmental Research. 2008;**2**(4):425-430, ISSN: 1735-6865

[24] Schneider P, Le Hung A, Wagner J, Reichenbach J, Hebner A. Solid waste management in Ho Chi Minh City, Vietnam: Moving towards a circular economy? Sustainability. 2017;**9**:286. DOI: 10.3390/su9020286.

[25] Schneider P, Oswald K-D, Littman R, Weiß B. Komplexes Abfallmanagement in Nordzypern. In: Tagungsband zur 11. Deponiefachtagung Leipzig Planung, Bau, Betrieb und Schließung von Deponien; 4 March 2015. pp. 31-45

[26] Ciuta S, Apostol T, Rusu V. Urban and rural MSW stream characterization for separate collection improvement. Sustainability. 2015;**7**:916-931. DOI: 10.3390/su7010916

[27] Apostol L, Mihai F-C. Rural waste management: Challenges and issues in Romania. Present Environment and Sustainable Development. 2012;**6**(2):105-114

[28] Wilson DC, Velis C, Cheeseman C. Role of informal sector recycling in waste management in developing countries. Habitat International. 2006;**30**(2006):797-808

[29] Shah R, Sharma US, Tiwari A. Sustainable solid waste management in rural areas. International Journal of Theoretical & Applied Sciences. 2012;**4**(2):72-75

[30] Mohammadi A, Amouei A, Asgharnia H, Fallah H, Ghanami Z. A survey on the rural solid wastes characteristics in North Iran (Babol). Universal Journal of Environmental Research and Technology. 2012;**2**(3):149-153. Available Online at: www.environmentaljournal.org

[31] Weng X. The Rural Informal Economy—Understanding Drivers and Livelihood Impacts in Agriculture, Timber and Mining. IIED Working Paper. London: IIED; 2015. Available from: http://pubs.iied.org/16590IIED, ISBN 978-1-78431-172-8

[32] UNICEF. The State of the World's Children. 2008. p. 134. Available online: https://www.unicef.org/sowc08/

[33] Solid Waste Management Technical Support Center (SWMTSC). Solid Waste Management Baseline Study of 60 New Municipalities. Lalitpur, Nepal: Engineering Study & Research Centre (P) Ltd.; 2017

[34] National Environmental Agency of Germany. Webpage of Sulakshana Mahajan. 2014

[35] Obersteiner G, Linzner RP, Scherhaufer S, Schmied E. Formalisation of informal sector activities in collection and transboundary shipment of waste in and to CEE - Introduction to the project "Transwaste". In: International CARE Electronics Office (Hrsg.): Going Green CARE INNOVATION 2010: From Legal Compliance to Energy-efficient Products and Services, 8th International Symposium and Environmental Exhibition; 8-11 November 2010; Wien. Vienna, Austria: International CARE Electronics Office Abstract Book, Abstract, 17, Full paper on CD

[36] István Z, Sándor RB, Négyesi B. Situation of informal waste picking in Hungary. In: International CARE Electronics Office (Hrsg.): Going Green. CARE INNOVATION 2010: From Legal Compliance to Energy-efficient Products and Services, 8th International Symposium and Environmental Exhibition; 8-11 November 2010; Wien. Vienna, Austria: International CARE Electronics Office Abstract Book, Abstract, 27, Full paper on CD

[37] Obersteiner G, Scherhaufer S, Linzner R, Pertl A, Schmied E. TransWaste - Status der wissenschaftlichen Analysen. In: ÖWAV (Hrsg.): "Wie viel Abfall braucht Österreich", Österreichische Abfallwirtschaftstagung; 4 und 5 Mai 2011; Graz. Österreichischer Wasser und Abfallwirtschaftsverband, Wien, ISBN 978-3-902810-07-6

[38] Tydlitatova EM, Havrland B, Ivanova T. Social awareness on waste production in rural areas. In: Engineering for Rural Development, Jelgava. 29-30 May 2014. pp. 560-564. Available online: tf.llu.lv/conference/proceedings2014/Papers/95_Mrlikova_E.pdf

[39] Aljaradin M, Persson KM, Sood E. The role of informal sector in waste management, a case study; Tafila-Jordan. Resources and Environment. 2015;**5**(1):9-14. DOI: 10.5923/j.re.20150501.02

[40] Karak T, Bhagat RM, Bhattacharyya P. Municipal Solid Waste Generation, Composition, and Management: The World Scenario, Critical Reviews in Environmental Science and Technology. 42:15. UK: Taylor & Francis Online; 2012. 1509-1630

[41] Paul T-G, Quetzalli A-V, Sara O-B, Carolina A. Waste characterization and waste management perception in rural communities in Mexico: A case study. Environmental Engineering and Management Journal. 2011;**10**(11):1751-1759. Technical University of Iasi, Romania. http://omicron.ch.tuiasi.ro/EEMJ/

[42] Buenrostro DO, Israde AI. La gestio´n de los residuos solidos municipales en la cuenca de Cuitzeo. Mexico Revista Internacional de Contaminacion Ambiental. 2003;**19**(4):161-169

[43] Castrejón-Godínez ML, Sánchez-Salinas E, Rodríguez A, Ortiz-Hernández ML. Analysis of solid waste management and greenhouse gas emissions in Mexico: A study case in the central region. Journal of Environmental Protection. 2015;**6**:146-159. http://dx.doi.org/10.4236/jep.2015.62017

[44] Sara O-B, Carolina AdV, Ramírez-Barreto ME. Formal and informal recovery of recyclables in Mexicali, Mexico: Handling alternatives. Resources, Conservation and Recycling. Engineering Institute of the Universidad Autónoma de Baja, CA, USA; 2001;**34**:273-288

[45] Medina M. Serving the unserved: Informal refuse collection in Mexico. Waste Management and Research. 2005;**23**:390-397. DOI: https://doi.org/10.1177/0734242X05057698

[46] El-Informador. Los diez productos para reciclar mejor pagados. 2015. Revised 7th May 2017. Available from: http://www.informador.com.mx/suplementos/2015/593073/6/los-diez-productos-para-reciclar-mejor-pagados.htm

[47] Department of Environmental Affairs. National Waste Information Baseline Report. Pretoria, South Africa: Department of Environmental Affairs; 2012

[48] Asian Development Bank (ADB). Solid Waste Management in Nepal: Current Status and Policy Recommendations. Mandaluyong City, Philippines: Asian Development Bank; 2013. ISBN 978-92-9254-232-0

[49] Asian Development Bank (ADB). Status of Solid Waste Management in 58 Municipalities of Nepal, Technical Assistance Consultant's Report: IPE Global Private Limited in association with Environment Resource Management Consultants (ERMC) (P.) Ltd. and Full Bright Consultancy (Pvt.) Ltd., Lucknow, Uttar Pradesh, India. Project Number: 44069. 2012

[50] Nhu Quynh D, Sakanishi K, Nakagoshi N, Fujimoto S. Tomoaki Minowa Potential for rice straw ethanol production in the Mekong Delta, Vietnam. Renewable Energy. 2015;**74**(2015):456-463. DOI: http://dx.doi.org/10.1016/j.renene.2014.08.0510960-1481/© 2014 Elsevier Ltd

[51] Scherhaufer S, Linzer R, Obersteiner G, Schmied E, Pertl A, Kabosch U. Trans Waste Formalisierungsidee 1: Integration des informellen Sektors in den Aufbau von Re-use und Repair-Netzwerken in Kooperation mit sozialwirtschaftlichen Integrationsunternehmen. In: ÖWAV (Hrsg.): "Wie viel Abfall braucht Österreich", Österreichische Abfallwirtschaftstagung 2011; 4 und 5 Mai 2011; Graz. Österreichischer Wasser und Abfallwirtschaftsverband, Wien, ISBN 978-3-902810-07-6

Chapter 4

Decentralized Composting of Organic Waste in a European Rural Region: A Case Study in Allariz (Galicia, Spain)

Iria Villar Comesaña, David Alves, Salustiano Mato,
Xosé Manuel Romero and Bernardo Varela

Additional information is available at the end of the chapter

http://dx.doi.org/10.5772/intechopen.69555

Abstract

The inclusion of sustainability and circular economy principles, as well as the compliance of the European requirements in municipal waste management, involves improving the waste separation, recovery and valorization. The current municipal solid waste management system of Galicia (Northwestern Spain) that includes most of the municipalities involves the treatment of biowaste (mixed in the same container with the nonorganic rest fraction) in a single management facility. This biodegradable fraction, which accounts for 42% of the total amount of household waste, is treated by incineration for energy recovery. The local government of Allariz (Galicia) undertook a project to implement a management model decentralized for biowaste separation and treatment through composting. Municipality structure (type of housing, urban and rural areas, etc.) made it necessary to implement different composting systems: home composters, community composting islands and a dynamic composter. During the first year of start-up of the management model, the level of citizen acceptance was adequate, biowaste was correctly segregated and good quality compost for soil fertilizer was obtained. So, a reduction of around 8% of the mixed waste sent to the centralized treatment facility was observed. The biowaste recovery had also resulted in a recycling improvement of all remainder fractions.

Keywords: compost, organic fraction of municipal waste, circular economy, recycling, citizen participation

1. Introduction

The management and planning of the municipal solid waste produced by citizens in their dwellings is one of the main objectives to be addressed by the circular economy principles. The European Union, in its commitment to the environment and sustainable development, promotes among its members the implementation of concrete measures and actions in order to improve current conditions and establish a legal framework for the proper management of municipal solid waste. The European Parliament adopted the Directive 2008/98/EC on waste [1], laying down measures to protect the environment and human health by preventing or reducing the adverse impacts of the generation and management of waste and by reducing overall impacts of resource use and improving the efficiency of such use.

In densely populated and urbanized areas of the European continent, there are many alternatives for the management of this type of waste [2–4] but also in other continents, where the population density is much greater and the establishment of urgent measures becomes an essential work to avoid negative impacts on the environment or human health from household waste [5–7]. But it is also necessary to carry out actions in rural or semirural areas where the management of household waste must be adapted according to the needs of each area so that the objectives established by the regulations can be achieved in viable environmental and economic conditions. In Spain, 23% of the population lives in rural areas, according to the resident population, and the particularities of each zone make it difficult to manage municipal waste correctly [8].

The Spanish law 22/2011 on waste and contaminated soil [9], which transposes European Directive 2008/98/EC, sets the target that, before 2020, the amount of domestic and commercial waste destined for the preparing for reuse and the recycling for paper, metal, glass, plastic, biowaste or other fractions shall be increased to a minimum of overall 50% by weight. This legislation establishes a waste hierarchy with the following order from highest priority to lowest:

- Prevention: a set of measures taken at the design, production, distribution and consumption stages of a substance, material or product, to reduce the quantity of waste, the adverse impacts of the generated waste on the environment and human health and the content of harmful substances in materials and products.
- Preparing for reuse: include checking, cleaning or repairing recovery operations, by which products or components of products that have become waste are prepared so that they can be reused without any other preprocessing.
- Recycling: any recovery operation by which waste materials are reprocessed into products, materials or substances whether for the original or other purposes. It includes the reprocessing of organic material but does not include energy recovery and the reprocessing into materials that are to be used as fuels or for backfilling operations
- Other recovery, including energy recovery, when it occurs with a certain level of energy efficiency.
- Disposal: any operation which is not recovery even where the operation has as a secondary consequence the reclamation of substances or energy

In order to achieve the targets established by the legislation, Spanish municipalities assume different management separate collection systems for household waste. These systems can be summarized in four models in municipalities with more than 50,000 inhabitants and in six models in smaller municipalities of 5000 to 50,000 inhabitants [10]. The Autonomous Community of Galicia, located in the northwest of Spain, has a total area of 29,574 km^2 and a population of 2,718,525 inhabitants in 2016. With a population density of 92.2 inhab/km^2, the most urbanized areas are mainly concentrated on the coast, while dispersed and rural population centers are established in the interior and the East of the Community. In Galicia, there are three different collection systems of municipal waste [11] that are summarized in **Figure 1** with the following characteristics:

- Model 1 of the Galician Society of the Environment S.A. (SOGAMA) to which are attached 295 municipalities and that covers 82.5% of the population of the community. Around 805,355 tonnes of waste were managed in the year 2015.
- Model 2 of the Treatment of Urban Waste of A Coruña to which are attached 10 municipalities representing around 14.3% of the population of Galicia. Around 174,318 tonnes of waste were managed in the year 2015.
- Model 3 of Sierra of Barbanza Environmental Complex to which nine municipalities belong, which represents around 3.1% of the Galician population. Around 32,220 tonnes of waste were managed during the year 2015.

All Galician municipalities are adhered to the models presented in **Figure 1** for municipal waste management, and there are no illegal landfills for the fractions considered in this study. The three management models implement containers in the public road for the municipal waste collection where the citizens deposit the waste generated in their homes in different fractions. The three models have independent containers for the separate collection of glass

Figure 1. Models for the collection of household waste of the Autonomous Community of Galicia (source: prepared by the authors based on the information from Urban Waste Management Plan of Galicia [11]).

and paper-cardboard, but present substantial differences in the other two containers. In the model 1 (SOGAMA model), there is no differentiated collection of the organic fraction, while in the other two models, there is a differentiated collection of biowaste. The models 2 and 3 have a yellow container for the inorganic fraction where lightweight packaging and the rest of wastes that do not present differentiated separation are deposited and another container for the separate collection of the organic fraction. In the system 1, at the majority of citizens of the community disposal, there is a yellow container where the lightweight packaging (plastic, metal and liquid packaging board) is deposited and another container for the rest or mixed fraction, i.e., all wastes that are not subject to separate collection, in this case, biowaste along with sanitary textiles, ceramic waste, household cleaning waste, etc.

As indicated above, most municipalities of Galicia (82.5% of the population) are included under a centralized municipal waste collection system. So, waste can travel more than 150 km before proceeding to its management in the treatment plant. To SOGAMA's facility arrives the waste deposited in the yellow container and the mixed container (organic and nonrecyclable waste). The materials collected in the lightweight packaging container are classified according to its different typology, and later, they are sent to the recycling centers to be transformed into new products. The waste collected in the mixed container, once separated the materials that can be recycled (steel and aluminum fundamentally), is subjected to an energy recovery process, whereby they are incinerated to produce energy. The biodegradable fraction, which accounts for about 42% of the total amount of wastes generated in housing (**Figure 2**), is not collected in a differentiated way and is segregated together with the wastes deposited in the

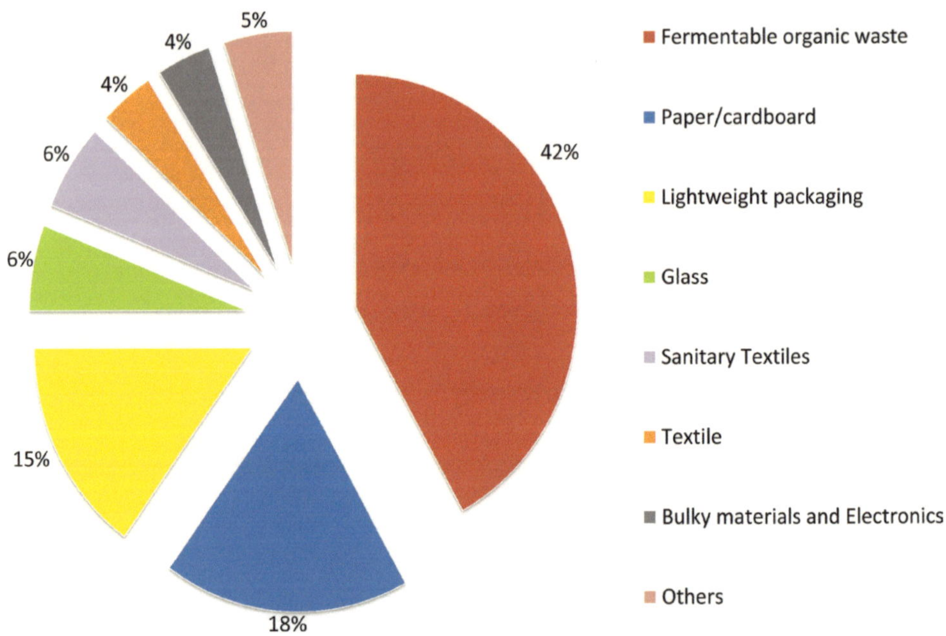

Figure 2. Composition of the household rubbish in Galicia [11].

mixed container. Therefore, the end-use of the biowaste, once it reaches the treatment plant, is the incineration for energy recovery; however, these biowastes have a low calorific value due to their high water content.

The definition of biowaste is described in Spanish law 22/2011 [9] and includes biodegradable garden and park waste, food and kitchen waste from households, restaurants, caterers and retail premises and comparable waste from food processing plants. Both the implementation of sustainability and circular economy criteria in waste management, as well as ensure compliance with European requirements, mean improving the segregation of biowaste and its treatment. Composting is a simple, low-cost recovery technology that enables organic waste and by-products to be transformed into biologically stable materials called compost. The compost can be used as an amendment and/or soil fertilizer and as a substrate for plant growth, reducing the environmental impact of biowastes and making it possible to take advantage of the resources contained in them. Composting is defined as a controlled biooxidative process, which develops on heterogeneous organic substrates in solid state, due to the sequential activity of a great diversity of microorganisms present in the substrate. Under these assumptions, composting of the organic fraction of household waste is presented as an economically accessible and adequate option with the environmental requirements. For example, in India they have bet on the decentralization of composting for this type of waste in cities for several years ago [12]. For that, it is necessary to involve the population in waste separation and the implementation by local entities of composting systems. Finally, it is possible to close the cycle of organic matter by obtaining a product of good quality, compost, for local, community or individual use.

In this way, the start-up of new models of organic waste management in the municipalities through the use of composting is growing exponentially in the Galician community. Composting experiences are being currently carried out in San Sadurniño and municipalities of the province of Pontevedra included in the "Revitaliza" program [13].

2. Case study in Allariz

2.1. Study area

The municipality of Allariz is located in the province of Ourense belonging to the Autonomous Community of Galicia located in the northwest of Spain (**Figure 3**). Allariz has an area of 85.3 km^2 where a population of 5982 inhabitants distributed in 92 population centers and with a density of 70 inhab/km^2. According to the indicator of rurality established in Urban Waste Management Plan of Galicia [11] that takes into account, among others, data on population, population density, tourist and commercial level, the municipality of Allariz has a rural character. Allariz is located in the area of influence of the most populous municipality of the province, the urban municipality of Ourense, with approximately 100,000 inhabitants. The population structure of Allariz is distributed, from the interior to the exterior of the municipality, in the old town with housing of different heights, an area with new buildings of heights from three to four and, finally, a more rural area consisting of detached or semidetached houses with garden. In addition, there are also many shops in the town center and an industrial estate on the outskirts. The

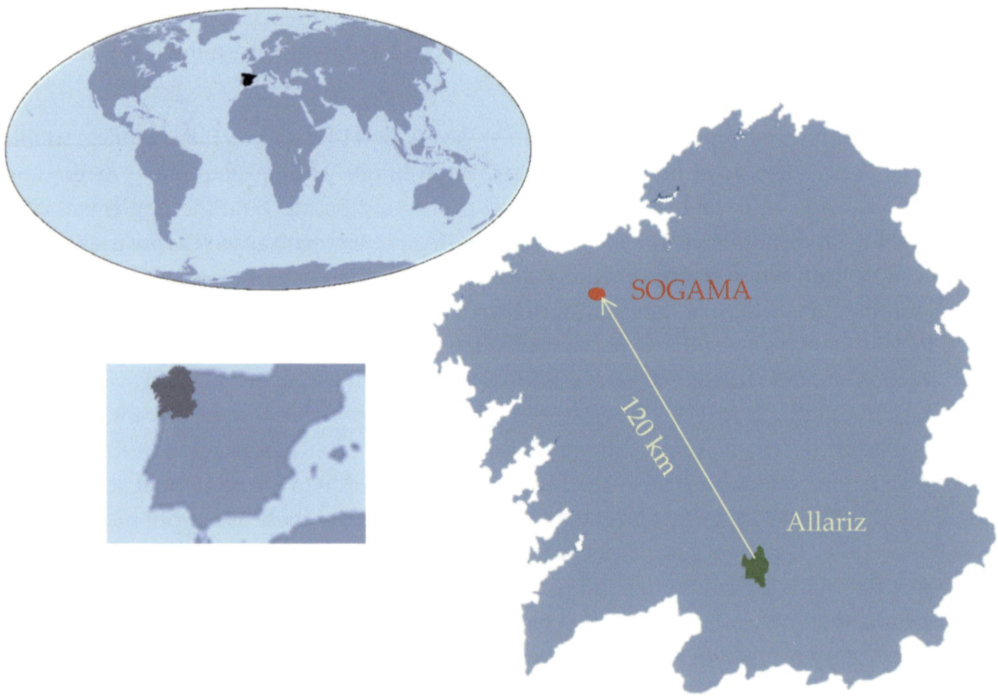

Figure 3. Location of the municipality of Allariz and the waste treatment plant of SOGAMA.

household waste of the mixed fraction and the packaging fraction must travel around 120 km away in a straight line (**Figure 3**).

At the end of 2014, Allariz set in motion to implement the separation, treatment and use of organic matter throughout the municipality. The following action lines were proposed:

- Promotion of single-family self-composting.
- Development and promotion of community composting.
- The specific collection of organic matter coming from big waste producers, such as catering establishments and food companies, for their joint composting.

These action lines aim to increase the recycling rate and minimize the percentage of wastes that are segregated together to be incinerated. Increasing the recovery of biowaste implies diverting them from incineration by reducing the emissions of greenhouse gases emitted during their combustion and their transport. Composting of biowaste produces compost for private or municipal use that allows closing the circle of organic matter by returning to the soil the nutrients extracted by plants and animals during their growth and development. This also leads to a greater use of waste in accordance with the hierarchy imposed by European regulations, and an efficient waste management model capable of contributing to sustainable development is consolidated.

Indirectly, it is also expected that by improving the household waste separation there will be an improvement in the existing separate waste collection of the different fractions: packaging, paperboard and glass. Thanks to increased citizens' environmental awareness can be reduced economic management costs, which arise as a result of the collection, transport and treatment of waste outside the municipality, in addition to environmental costs. The internalization of the organic fraction management in the municipality itself allows the strengthening of employment in a rural region.

Another objective of the plan is to recover the green waste that is generated in the houses with garden and during the maintenance of the green areas of the municipality. This green fraction consists mainly of pruning remains, grass clippings and leaves that can be crushed to be used as a structuring agent in the composting of biowaste.

2.2. Methodology

2.2.1. Composting process

Composting is a controlled biooxidative process, in which a heterogeneous organic substrate undergoes a thermophilic stage and a transient release of phytotoxins, obtaining as products: carbon dioxide, water, minerals and stabilized organic matter called compost [14]. Due to the high microbial activity during the composting process, the temperature increases and accelerates the degradation and mineralization of the organic matter. Changes in temperature throughout the process allow differentiating four phases [15]:

- Mesophilic phase: characterized by the increase of the temperature from values close to the ambient temperature until reaching approximately 45°C. During this phase, the mesophilic microorganisms begin to slowly degrade the organic matter.
- Thermophilic phase: at temperatures above 40°C, the mesophilic activity drops, and the degradation begins a thermophilic stage, reaching values of 60–70°C. The thermophilic phase is very important, since reaching temperatures of this magnitude produces pasteurization of the product, destroying the pathogenic microorganisms and seeds of invasive plant species, so this ensures the hygienization of the material produced. From the 80°C, excessive heat can cause the death of most microorganisms to stop the degradation activity, so the composting must be controlled so that these temperatures are not exceeded.
- Cooling phase: the mixture begins to cool because easily degradable materials have been consumed during the mesophilic phase and mainly in the thermophilic phase. As a consequence of this, a return to the mesophilic stage occurs and the temperature drops to near the values of the ambient temperature.
- Maturation phase: at this stage, complex secondary condensation and polymerization reactions occur, which results to the compost as final product. It is necessary that this phase has the duration such that the material acquires the maturity and the necessary stability of an organic amendment of agricultural application.

2.2.2. *Composting systems*

Thus, the implementation of the biowaste management model through composting was carried out at three levels:

- Individual composters for single-family houses. A total of 220 family composters were given voluntarily to the residents and with the only obligation to use the composter bin in order to treat the organic household waste to obtain compost. These individual composters have a volume of 300 L and, mainly, were delivered to homes away from community composters. The methodology used in these composters basically consists of alternating layers of biowaste with layers formed by leaf litter, crushed grass, chip or shredded pruning wastes that can be obtained in the garden of the participants themselves. In this system, composter users and those responsible for obtaining the compost are the family that generates the biowaste, and the municipality staff undertakes to carry out training and questioning by users.
- Community composters for residential areas and buildings of various heights. The municipality of Allariz implemented a total of 24 composting islands (areas for the placement of a group of composters) located on public land in urban and periurban areas with more than 130 composters. Each composter has 1000 L of capacity and is considered to accept the organic matter deposited by approximately 15 families. These modular composters are made of recycled plastic slats that enable the walls to be completely removable and have the advantage of work as independent or dependent modules of composting, making easier the movement and transfer of compost. Each island is constituted up of a number of modules that depend on the volume of population they serve. Bulking agent used in this system is shredded wood that comes from a biomass company located in the industrial estate of the municipality and the green waste provided by neighbors and municipal services. The work of the neighbors/participants focuses on separating correctly the organic fraction generated in their dwellings, getting a waste free of improper materials such as plastics, metals or glass, and transferring the biowaste to the nearest composting island. The town council distributed 10 L cubes to the neighbors to facilitate the transfer of biowaste. The organic fraction deposited in the composters, called the contribution composters, must be covered with the bulking agent, present in all the islands, to avoid the appearance of insects and odors. The other composters of the island, other than those of contribution, are used in the maturation of compost. The responsibility of the municipal staff is to carry out the works of mixing, irrigation, screening, turning and transferring between composters of the material, as well as to distribute the compost to the neighbors who have participated in this initiative.
- Dynamic composter for biowaste produced by big waste producers. This dynamic bioreactor Big Hanna model T120 consists of a rotating cylinder orientated horizontally with temperature sensors at different positions along the cylinder and a continuous aeration with fan. Biowaste is fed through a screw conveyor in the hopper situated in the front of the composter, rotation of the cylinder moves the material and the compost is emptied through the back side after about 8 weeks. The bioreactor can accept a load of 300–500 kg per week and has a capacity of about 3 m^3. In this electric bioreactor, the periodic rotation of the cylinder itself produces the material mixture and helps the aeration of the material accelerating the decomposition of the organic matter. The municipal staff collects the organic matter generated

http://dx.doi.org/10.5772/intechopen.69555

and separated by the catering establishments and food companies to be transferred to the bioreactor. The bioreactor is designed to use pellets as a control material for excess moisture produced during the process and is added together with the biowaste in the hopper by the municipal staff. After the composting cycle, the pellets become part of the final compost. The compost generated in the bioreactor is sieved by a 1-cm mesh and transferred to a composter until it reaches the parameters of maturity and stability enough to be applied to the soil with quality assurances.

The planning, distribution and start-up of the model were carried out along with information, training and awareness campaigns to the neighbors and the different parts involved in the model. The educational work of the citizens gave appropriate answers to the doubts and questions asked by the users. On the other hand, the choice of the message conditions the content, which must always be accurate and verified, and also transmitted with simplicity and clarity to achieve the objectives of the campaign. Knowing the municipality, its territory and its idiosyncrasy, helps in understanding the demands and terms of the message to be transmitted. **Figure 4** shows the poster that is present in the community composting islands of Allariz. With this poster, it is intended to draw the attention of neighbors, but also of tourists and visitors, increasing the initiative's visibility.

Therefore, the development of this management model was carried out jointly with the activities of dissemination and education through training and awareness talks to the citizens of Allariz, in both the rural area and the old town. Accordingly, it is worth noting the activities carried out for children in schools taking advantage of the cross-curricular thematic of waste management in school curricula. In addition, door-to-door visits were made to inform about the advantages of composting and 600 cubes of 10 L of capacity were delivered to the neighbors to transport their biowaste to the community composters and to the individual composter. For the composting of the biowaste produced by the big waste producers, 20 containers of 120 L of capacity were distributed and a schedule of door-to-door collection was established.

a) b)

Figure 4. (a) Community composting island with an area of green waste contribution on the left and composters on the right. (b) Detail of one of the informative posters.

Training activities and follow-up work on the individual composters were carried out by the Association for the Ecological Defense of Galicia (ADEGA). The staff of the Ramón González Ferreiro Foundation, together with the municipal staff of Allariz, monitor daily (from Monday to Friday) the community composting islands, performing data collection works on temperature, moisture, type of material contributed, etc., and also place informative panels on these islands on the inappropriate uses that are detected. The implementation of corrective measures of problems encountered, such as excessive grass feed, presence of thick pruning remains or improper inputs, is essential for participants to learn in the most appropriate way and obtain good quality compost without causing odors, insect problems or other annoyances.

The Environmental Biotechnology group of the University of Vigo carried out the sampling and analysis of compost, both from the community composters and from the bioreactor for the biowaste of the big producers. In total, 29 samples of compost were analyzed and 5 characterizations of the bioreactor input material were performed.

Once the samples were analyzed and the data obtained were evaluated, the compost was distributed to the participating neighbors. These types of events are an important part of the training of citizens because the delivery of the final compost is valued as an award for participation and volunteer work. At this time of learning, citizens who do not have the opportunity to do compost in their own homes see as their work of separation and transfer of a substance, from which they are going to be undone because it has no value, is transformed into a product that can be used in flowerpots or urban gardens. The concept "from waste to compost" helps to a greater implication of the citizens in the model and greater awareness of environmental care with a change of mentality toward a perception of waste to resource, which leads to an improvement in the selective collection of all fractions and the quality of these fractions.

2.2.3. Bulking agent

The need to mix biowaste with a material that provides porosity, in order to facilitate the aerobic conditions for the composting process, becomes the key factor of the bulking agents. However, porosity is not the only intrinsic property required of a good bulking agent, but also, inter alia, its ability to capture and/or cede water according to the needs of the process [16]. The complexity and heterogeneity of the biowaste deposited by the neighbors involves a great variability in the moisture values. Thus, while the remains of fruit and vegetables are high in water, the bread, eggshell or fish remains have a lower content. The initial moisture content of biowaste is around 70–80% [17]. However, the continuous contribution of biowaste by citizens causes materials with different degree of degradation to be mixed inside the composter, causing variations in the density and humidity of the material. The physical, chemical and microbiological characteristics of the bulking agent are determinant, and all of them will influence, to a greater or lesser extent, the composting process. In this way, it is necessary to take into account the characteristics of the bulking agent, using the appropriate machinery, tools and handling, which provides the best conditions of this key factor.

The type of bulking varies depending on the composting system used. In the individual composters, biowaste is supplied by layers and is covered by green wastes from the dwelling of the participants: leaves, straw or shredded pruning. As for the bioreactor, it is designed to use pellet as a moisture control material. For community composters, mainly crushed wood is used as bulking. This material is represented mainly by a particle size of 1–2 cm with an average of 43% (**Figure 5**). This size stands out above the others because, on the one hand, it provides a greater porosity to the mixture without damaging the increase in temperature and, on the other hand, a considerable part of this fraction is recovered for the next composting cycle. The particle size of the buking agent must be balanced. The thicker fractions have a higher percentage of recovery, so that when sifting the compost this fraction is recovered and can be recirculated for a new cycle of composting. However, particle sizes greater than 2 cm increase to a large extent the presence of macropores in the biowaste, which, if their percentage in the mixture is very high, can cause a decrease in temperature by an excess of aeration. The compost is usually sieved around 1 cm, so that bulking fractions smaller than this size are not recovered in the screening process and become part of the compost. In addition, an excess of small particles can fill the pores and complicate the aeration of the waste during the process.

To size the needs of the bulking agent, the particle size distribution provides very important information. In those moments in which there are difficulties in getting bulking agent, it is necessary to pay attention to the recovery rates of this one because this material degrades in the process and the finer particle sizes are not recovered, once it passes through the sieve, to recirculate and to use in another composting cycle.

On the other hand, it is convenient to deposit the bulking agent in a box protected from rain in order to control the moisture content of the compost by irrigation in all the three systems.

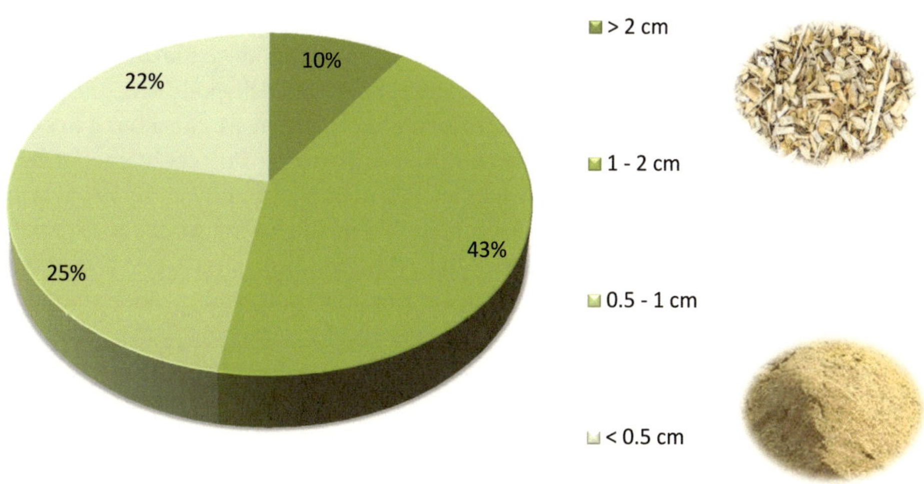

Figure 5. Particle size distribution of the bulking agent used in community composters (source: prepared by the authors based on the results of this study).

An excess of moisture during composting can produce bad smells due to the saturating the mixture. A pronounced decrease in moisture, as a consequence of the loss of water produced by the elevated temperatures inside the composter [18], must be compensated by irrigation in order not to stop the degradation of organic matter. In community composters, the green waste used for neighbors to cover biowaste has a greater effect the drier the bulking agent, reducing the odors released by them. During periods of time when municipal staff is not present to mix the bulking agent with biowaste, anaerobic episodes can occur due to the absence of porosity of the organic matter provided by the neighbors, in turn these materials may have several days in the cube or bag which may increase the presence of putrid odors resulting from anoxic decomposition. For this reason, a bulking agent capable of retaining these emissions, as long as the structure of the material is not corrected, can minimize odors and avoid possible annoyances to citizens

2.3. Results and discussion

2.3.1. Composting systems

In the three composting systems, periodic monitoring was carried out by members of the city council and collaborating entities in order to implement the changes in the management model and to control the composting process of biowaste. **Figure 6** shows a small summary of the versatility conditions presented by the three systems.

All the systems implemented in the decentralized model present several advantages depending on the needs of the user collective, being decisive variables such as the volume of waste production, the availability of an appropriate place for the development of the system, the necessary resources and, fundamentally, the economic cost that each system has.

The frequency with which the household waste is deposited varies in relation to the distance that the neighbors have to travel to deposit their wastes [19]. Thus, in the rural areas, the presence of individual composters is common due, among other factors, to the distance separating houses from containers. Other factors such as climatology, waste disposal schedule, reduction of biowaste and the need for a fertilizer for the orchard or garden are considered to be advantages of individual composting. At the level of the composting process, it should be noted that unlike individual composters, the modular system of community composters facilitates the emptying and transfer of the composted material, thanks to the simple manipulation of the plastic slats. At the same time, the use of this recycled plastic involves a reduction of up to 52% in environmental loads due to savings in raw materials, energy and emissions [17]. Another advantage that presents this modular system of composting is the activator effect that provides the proximity of the composters. This is because the proximity of a module with thermophilic temperatures facilitates the activation of the composting process in the nearest modules, reducing the effect caused by the low ambient temperatures that can occur at night or in colder seasons.

The effect created by the volume of material to be treated also has consequences in the process. The individual composters, with smaller volume, are more influenced by ambient temperatures than the material present in the community composters and, above all, the material of the biore-

	Individual composter	Community composter	Big Hanna bioreactor

Does the bio-waste leave the home?	No	Yes	Yes
Does the material reach thermophilic temperatures?	Lacks continuity	Yes	Yes
Is there a continuous contribution?	No	Yes	Yes
Is a collection schedule necessary?	No	No	Yes
Municipal management	No	Yes	Yes
Economic cost	Low	Medium	High
Level of physical effort	Low	High	Medium

Figure 6. Comparison of the three composting systems set up in the municipality of Allariz: individual composter, community composter and bioreactor for large producers.

actor that presents greater volume and isolation with the exterior. In addition to these reasons, the frequency of biowaste inputs to an individual composter depends to a large extent on the size of the family residing in the dwelling, i.e., small families of two or three members generate less volume of biowaste to feed the composter. Based on these premises, the individual composting lacks the intensity of the other two systems and, consequently, more time is necessary to produce compost. It is considered that the addition of meat and fish remains can accelerate the composting process [20].

The compost that is obtained in the bioreactor still lacks a state of maturity and stability required to be applied to the soil [21], in a system as a composter or through another system, to reach the conditions established by the legislation, reason being additional space is required for maturation [22]. In addition, the bioreactor system presents greater economic cost due to the high investment in the equipment purchase. To this must be added the cost for the collection and transfer of the organic waste, which is carried out in a determined schedule, and the cost of personnel for loading, unloading and control of the process.

2.3.2. *Community composters*

To evaluate the quality of compost generated in the community composting islands, these were analyzed before being delivered to the participants. A summary of the data of the 19 composts sampled in January 2017 are shown in **Table 1**, which also shows the variability of the samples on the general characteristics established by the legislation on fertilizer products [23] and other important parameters of stability and maturation [24, 25]:

- Compacted bulk density. Compost presented an important variability in this parameter and was betwefact that this parameter is affected mainly by the moisture and the distribution of the particles. As the organic matter degrades, the number of smaller particles increases, causing an increase in the bulk density. The higher the density, the lower the capacity to maintain adequate porosity values and higher compaction, although very low densities indicate numerous air spaces that can make water difficult for plants. The presence of particles of bulking agent in some composts caused the density decrease while the high moisture detected in several of them allowed higher densities.

- Stones and others inert materials. Both inert materials and stones are small-sized materials that remain at the end of the composting process and cannot be separated by refinement. 100% of the analyzed samples had a percentage less than 0.1% of inert materials greater than 2 mm. This is due, mainly, to the good separation of the wastes by the participants, which allows obtaining a very pure organic fraction and with a low presence of improper ones. In addition, the content of stones greater than 5 mm did not exceed 0.1%. The community composting system allows that once detected certain errors in the separation, these can be corrected through different mechanisms such as meetings with neighbors, information talks, e-mail messages or simply by direct contact with the participants.

- pH. At the beginning of the process, the biowaste deposited by the neighbors has a normally acidic pH due to its high water content and easily degradable organic matter. This measure is usually corrected until neutral and slightly alkaline values are reached in the final compost. With regard to the application of compost, considering that in Galicia most soils and water are acidic, it is advisable that organic amendments have a slightly basic pH to correct soil acidity and improve crop growth. All samples exceed pH 7, and one sample is above 8.

- Electrical conductivity. The evolution of this parameter is very important for final application of the compost to the soil since a high content can cause adverse effects on the germination and the growth of the plants. As the transformation of waste into compost progresses, salts

	Mean	S	Median	Percentile 2.5	Percentile 97.5
Inert materials > 2 mm (% sms)	<0.1	–	–	–	–
Stones > 5 mm (% sms)	<0.1	–	–	–	–
Bulk density compacted (g L^{-1})	396.70	104.02	376.52	254.64	577.50
Moisture (%)	57.67	13.15	61.15	33.82	75.12
Organic matter (%)	61.79	15.23	66.63	26.10	75.71
pH	7.66	0.26	7.70	7.17	8.04
Electrical conductivity (mS cm^{-1})	1.21	0.51	1.18	0.47	2.16
Total carbon (%)	29.20	8.31	34.05	11.10	36.12
Total nitrogen (%)	2.39	0.71	2.45	1.04	3.25
C/N ratio	12.25	1.28	12.00	10.14	13.90
N-NH_4^+ (mg kg^{-1})	163.3	389.8	77.1	28.3	1029.6
N-NO_3^- + N-NO_2^- (mg kg^{-1})	55.4	26.6	51.5	21.8	109.1
CaO (%)	6.88	3.24	6.40	2.00	13.68
K_2O (%)	1.73	0.68	1.81	0.67	2.67
MgO (%)	0.50	0.08	0.51	0.33	0.60
P_2O_5 (%)	1.43	0.51	1.29	0.52	2.17
SO_3 (%)	0.71	0.25	0.76	0.24	1.07
FeO (%)	0.41	0.20	0.35	0.20	0.83
Co (mg kg^{-1})	2.85	1.43	2.78	0.99	5.41
Mn (mg kg^{-1})	185.90	68.28	172.66	101.28	319.80
Mo (mg kg^{-1})	2.05	0.93	1.84	0.97	3.91
Germination index (%)	86.5	17.1	89.2	60.4	115.2
Maturation degree	IV–V	–	–	–	–
Salmonella spp (in 25 g)	Absence	–	–	–	–
Escherichia coli (ufc/g)	<1000	–	–	–	–

Table 1. Maturity and stability parameters in compost from community composters (source: prepared by the authors based on the results of this study).

and mineral components with different solubility are released. Irrigation during the process can reduce the content of these through the leachate. More than 60% of the samples had values higher than 1 dS m^{-1}, although only two samples exceeded values of 2 dS m^{-1}. Therefore, in this parameter, the use of compost does not present a risk to plant development.

- Organic matter. During the composting process, the intense microbial activity and high temperatures cause a reduction of the organic matter, in more or less proportion. Taking into account that compost is a product to be applied as an organic amendment, it is assessed that it must have a significant content in organic matter, above 40% [23], but that the organic matter is sufficiently stable. In this way, the analysis of parameters indicative of stability such as the self-heating test and the C/N ratio are performed. The self-heating test measures the heat released during microbial activity by classifying the material into five classes according to their maturity. All composts were classified according to their stability in classes IV and V indicative of mature compost. The values of the C/N ratio were all lower than 20 which are considered an adequate ratio for compost, although values below 12 are considered preferable [25], presenting 50% of the compost values lower than 12. According to the TMECC [24], the C/N ratio of compost is not an independent indicator of stability or maturity, so other indicators such as respirometry, pH, bulk density, organic matter reduction and self-heating must be considered.

- Another important value of maturity and stability is the germination index that is calculated by germination and root length of seeds growing in aqueous extracts of compost. Values higher than 80% are indicative of mature compost and the absence of phytotoxic compounds for plant growth being values below 50% indicative of immature compost [26, 27]. More than 60% of the samples reached values above 80%, which demonstrated a high degree of maturity, and none of the samples presented values below 50%.

- The content of pathogens was in accordance with the parameters proposed by the legislation and without exceeding the maximum levels of microorganisms. This is mainly due to the high temperature values reached during composting [20]. By means of the taking of temperatures, an evolution is detected according to the process of composting, reaching values that are around 60°C during the thermophilic phase. There are also sharp falls in temperature during the thermophilic phase due to occasional drops in the contribution of biowaste from the neighbors or sharp declines in the ambient temperature. Other factors that can affect the temperature are an excess of ventilation in the composters, although the composters that are closed can enter air of the outside, or also rainwater entries retained in the cap that enters the composter when the neighbors open it to deposit the waste. In addition, temperature oscillation is common throughout the maturation in community composters, which is due to the reactivation of the material by turning and homogenizing it. In this process, the material closest to the walls of the composter is usually wetter and less decomposed than the material inside the mass, and mixing these materials can trigger an increase in activity with the consequent increase in temperature. In addition, the increase of the ambient temperature or the increase of the solar exposition of the composters can reactivate the interior temperature of the material.

- Ammoniacal nitrogen. A single sample had higher ammonium values than those considered suitable for compost, 400 mg kg^{-1} [14], hence the high standard deviation indicated in the **Table 1**. The high ammonium content could be a consequence of problems of degradation of the organic matter during the composting process due to a lack of moisture in this sample.
- Total nutrient contents provide a measure of compost fertilization potential and, however, do not allow determining the bioavailability of these elements for the growth of plants and microorganisms living in the soil. Compost produced from biowaste presents a high proportion of reserve nutrients such as P, Mg, K and Ca, these macronutrients can reach values higher than other substrates, such as peat, and contain the amounts necessary for the growth of plants [28].

The Spanish legislation on compost, Royal Decree 506/2013 of 28 June on fertilizers [23], classifies compost into three categories according to the heavy metal content: classes A, B and C. **Figure 7** provides information on the content in heavy metals, indicating the frequency of samples corresponding to class A, B or C. All samples have a concentration below the class A threshold for cadmium, copper, lead and mercury with concentrations below 50% of the class A limit in copper, lead and mercury. For zinc, the concentrations detected in more than 70% of the samples meet the threshold of class A and the remaining ones are classified within class B. On the con-

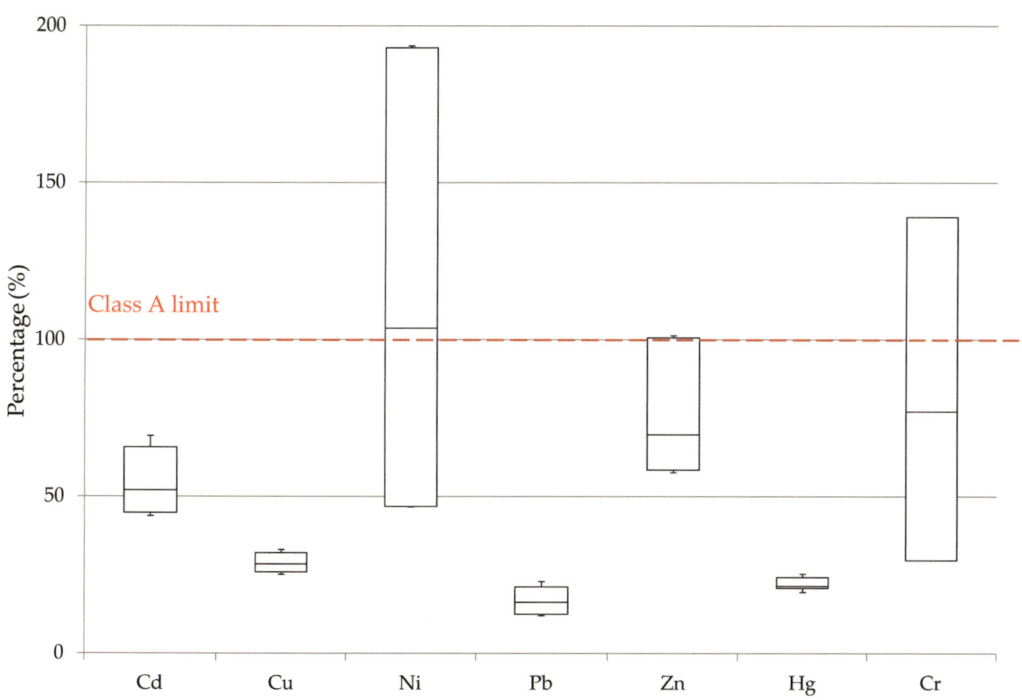

Figure 7. Boxplot with the heavy metals analyzed in the 19 compost samples generated in the community composters. Red line indicates the corresponding 100% with class A of compost.

trary, the samples show very high concentrations for chromium (37% of the samples are class B) and above all for nickel (53% of the samples are class B). The presence of these heavy metals in the final compost may have different sources, such as the presence of improper materials as elements outside the organic fraction of municipal waste selectively collected. On the other hand, Ansorena [17] concluded that the heavy metal content of the compost can be affected by the pollution of diverse exogenous sources and whose origin can be found in the auxiliary materials used, the environment, the process or the storage. The author shows as an example that in the composting plant of Lapatx (Guipuzkoa, Spain) high concentrations of nickel and chromium were detected and the analyses indicated that the material used as bulking agent contributed important amounts of these metals. Hence, more research is needed to find the source of nickel and chromium in the compost from community composters. Thus, compost obtained can be used without restriction for gardening and cultivation of fruit trees although its use should be valued for horticultural crops in compost of class B.

The compost sampled and analyzed are the first obtained after the start-up of the community composting system. In general, it is observed that the system allows the adequate management of the household biowaste, being necessary to determine the source of the contamination by heavy metals and to carry on with the continuous improvement of the system. In addition, as the population and municipal services assume more, experience and knowledge of the process will improve the system and quality of compost.

2.3.3. *Dynamic in-vessel composting*

Five characterizations of the input biowaste in the bioreactor were carried out, and nine samples of the output material of the bioreactor were analyzed. The material removed from the bioreactor during the first months of 2016 was sieved by a 1-cm mesh and matured in a composter taking samples at 2 months and 4 months of maturation. These values are represented in **Table 2**.

During the characterization of the input material of the bioreactor, it was observed that the percentage of improper ones is below 1% in fresh sample. The presence of plastic, metallic or glass wastes mixed with the input biowaste had a very low frequency, and their separation was carried out during the loading tasks in the bioreactor. This factor is important since the presence of improper ones causes pollutions in the organic material, so their absence makes possible that the levels of heavy metals classify the compost obtained by this system as fertilizer of class A, and therefore, in a compost without restriction of use. The low level of improper one indicates an adequate work of awareness of the big producer participants in the separation of biowaste. On the other hand, most biowastes introduced into the bioreactor, an average of 62%, were post-cooked wastes (leftover bread, pasta and vegetables), 26% were pre-cooked biowastes (peels, vegetables and fruits) and 12% were traces of paper napkins. The great presence of organic matter in a cooking process facilitates the biodegradation of the most resistant components.

During the composting process inside the bioreactor, there were high temperatures reaching 60°C because, as indicated in Ref. [29], the turnings significantly increase the duration of the thermophilic phase and, consequently, a greater degradation of the organic material. The turning of the material caused by the rotation of the drum facilitated the homogenization and the mixing

	T0	T1	T2
Inert materials > 2 mm (% sms)	<0.1	<0.1	<0.1
Stones > 5 mm (% sms)	<0.1	<0.1	<0.1
Bulk density compacted (g L^{-1})	385.72 ± 35.01	339.57 ± 32.1	468.22 ± 22.7
Moisture (%)	15.99 ± 0.8	31.20 ± 13.8	57.33 ± 10.7
Organic matter (%)	88.75 ± 2.2	84.53 ± 13.3	66.48 ± 9.1
pH	8.83 ± 0.5	8.24 ± 0.5	8.30 ± 0.5
Electrical conductivity ($mS.cm^{-1}$)	4.59 ± 0.7	2.52 ± 0.4	1.17 ± 0.2
Total carbon (%)	42.65 ± 0.4	41.1 ± 2.2	31.72 ± 1.9
Total nitrogen (%)	2.02 ± 0.4	2.54 ± 0.4	2.16 ± 0.2
C/N ratio	21.5 ± 4.2	16.18 ± 1.1	14.69 ± 0.9
N-NH_4^+ (mg kg^{-1})	1686 ± 465	504.4 ± 2.7	68.2 ± 2.8
N-NO_3^- + N-NO_2^- (mg kg^{-1})	100.3 ±2.5	94.6 ± 2.3	120 ± 5.6
CaO (%)	3.98 ± 0.85	5.33 ± 0.9	4.98 ± 0.3
K_2O (%)	1.09 ± 0.31	1.78 ± 0.6	0.97 ± 0.5
MgO (%)	0.32 ± 0.19	0.45 ± 0.2	0.38 ± 0.1
P_2O_5 (%)	0.88 ± 0.46	1.43 ± 0.4	1.24 ± 0.3
Cd (mg kg^{-1})	0.50 ± 0.2	0.47 ± 0.1	0.48 ± 0.1
Cr (mg kg^{-1})	8.61 ± 3.5	12.5 ± 1.8	35.6 ± 3.2
Cu (mg kg^{-1})	20.07 ± 15.9	21.7 ± 4.7	19.2 ± 3.5
Ni (mg kg^{-1})	4.46 ± 2.3	6.69 ± 2.7	17.50 ± 4.2
Pb (mg kg^{-1})	4.26 ± 3.72	<4.0	<4.0
Zn (mg kg^{-1})	88.37 ± 70.05	113 ± 34	112 ± 29
Hg (mg kg^{-1})	0.09 ± 0.02	0.06 ± 0.03	0.07 ± 0.02
Germination index (%)	76.99 ± 16.2	77.2 ± 1.5	87.45 ± 2.0
Maturation degree	II	IV	V
Salmonella spp (in 25 g)	Absence	Absence	Absence
Escherichia coli (ufc/g)	<10	<10	<10

T0 material after 8 weeks in the bioreactor, T1 compost maturated during 2 months, T2 compost maturated during 4 months.

Table 2. Maturity and stability parameters in compost from Big Hanna composter (source: prepared by the authors based on the results of this study).

of the material, achieving the elevation of the temperatures until lowering the moisture to values below 30%. These low values can cause water stress in microorganisms by slowing down the process [30] and inhibiting the degradation of biowaste. In order to obtain compost with better conditions of maturity and stability, the compost of the bioreactor output was matured in plastic composters.

The compacted bulk density of the compost at the exit of the bioreactor is low which may be due to the presence of pellets which are mixed at the inlet with the biowaste. The pellet absorbs, on the one hand, the excess water present in the biowaste and, on the other, the metabolic water produced during the degradation of the material, losing its structural stability [31]. This reduces the presence of leachates and bad odors during the process but causes a decrease in compost density.

The quality of the compost obtained is the result of an optimal separation of the waste by the establishments adhered to the program, which allowed obtaining a very pure organic fraction with a low presence of improper ones and other materials like stones. Hundred percent of the analyzed samples had a percentage less than 0.1% of impurities greater than 2 mm, and no stones larger than 5 mm were found.

The reduction of organic matter during the process leads to an increase in the concentration of some heavy metals (**Table 2**); however, the quality of the biowaste and the lack of external pollutions allow all the samples analyzed to have a concentration below the threshold of class A established by the Spanish regulations for Cd, Cr, Cu, Ni, Pb, Zn and Hg.

As in community composters, the presence of high temperatures during the thermophilic phase reduces the content of pathogens to values below the levels allowed by state legislation for both *Salmonella spp* and *Escherichia coli* levels.

Thus, the compost is a product free of pathogens and seeds, as a consequence of the pasteurization to which the waste is submitted inside the bioreactor, but that is unstable and self-heating when adding water and oxygenates it by turnings. In this way, the maturation process of the compost allows to improve the parameters of pH, electrical conductivity, C/N ratio, ammoniacal nitrogen, germination index and self-heating test. In 2 months, turnings and the moistening of the compost allow its stabilization and, after 4 months, the mineralization of the organic matter is improved, reaching optimum quality parameters. Therefore, the Big Hanna compost must be matured with mixing and irrigation for at least 2 months to obtain compost that meets the criteria of stability and maturity. In addition, the performance of maturation in composters or similar systems avoids cross-contamination and protects the material from drying and excessive leaching by precipitation.

2.3.4. Other waste fractions

The complete implementation of the decentralized model of Allariz (door-to-door collection of large producers, individual composting and community composting) was carried out during the spring of 2016. **Figure 8** and **9** shows the corresponding monthly data of the fractions of waste collected by the municipality according to the SOGAMA model implemented: glass fraction, paper-cardboard fraction, lightweight packaging fraction and mixed or rest

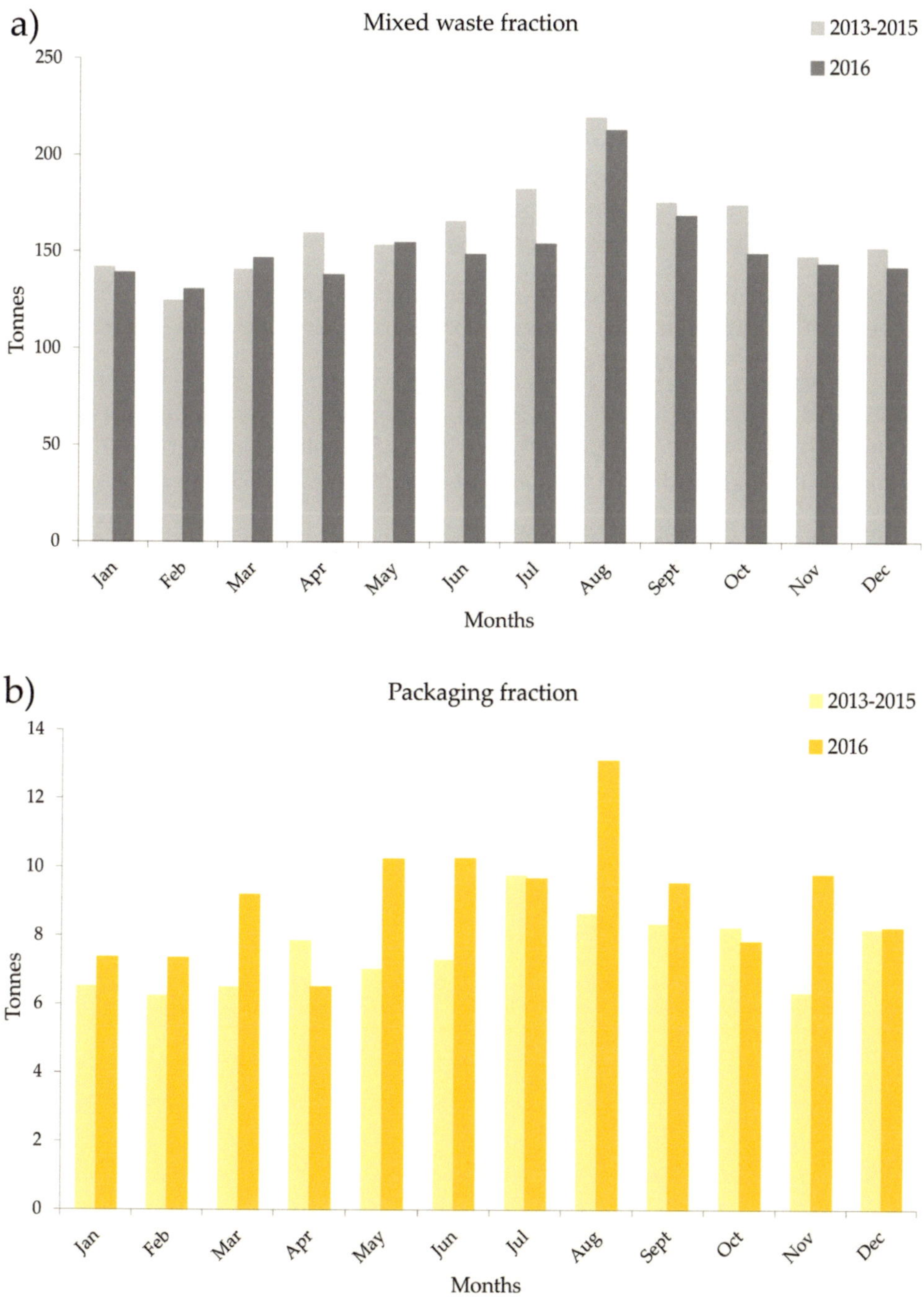

Figure 8. Different fractions collected in the municipality of Allariz: (a) average monthly production of the years 2013–2015 vs monthly production of the year 2016 of the mixed fraction, (b) average monthly production of the years 2013–2015 vs monthly production of the year 2016 of the lightweight packaging fraction.

fraction. It is noteworthy that the results show how Allariz is a rural municipality in which the presence of tourists and neighbors with second dwelling increases during the summer months, especially the month of August. The amount of waste collected during the summer period can be up to 40% higher than that collected in the months of lower production.

During the period beginning in the spring of 2016 until the end of the year, a reduction in the mixed fraction tonnes sent to SOGAMA by 7.3% over the average monthly tonnes delivered during the years 2013–2015 was observed (**Figure 8a**). The collection rate of mixed waste in the year 2016 was 0.838 kg/inhab/day. The reduction in the mixed fraction collected was accompanied by:

- An increase of 20.1% in the collection of the lightweight packaging fraction with respect to the years 2013–2015 (**Figure 8b**). The collection rate of lightweight packaging fraction in the year 2016 was 0.050 kg/inhab/day.
- An increase of 8.5% in the paper-cardboard fraction with respect to the years 2014–2015 (**Figure 9b**) (there was no selective collection of this fraction in containers in 2013). The collection rate of paper-cardboard fraction in the year 2016 was 0.045 kg/inhab/day.
- An increase in the collection of the glass fraction around 11.8% compared to 2015 (**Figure 9a**) (no data of years 2013 and 2014 are available). The collection rate of glass fraction in the year 2016 was 0.087 kg/inhab/day.

The collection of total waste during the year 2015 presented a rate of 1.038 kg/inhab/day, while the waste collection of 2016 corresponded with 1.019 kg/inhab/day. Therefore, the reduction in the rate of collected waste, namely the reduction in the mixed fraction, is due to two causes:

- The deviation of the organic fraction toward the three implemented composting systems. The neighbors and the big producers segregate the biowaste to destine them to composting and do not introduce them in the mixed fraction container. In the same way, the green waste generated by the neighbors (grass clippings, pruning and leaves) is deposited in the areas of contribution of the community composting islands for their use as bulking agent, so they are not introduced into the rest container.
- Improvement of recycling of other fractions that are segregated incorrectly in the mixed fraction by citizenship. Thus, it has been observed that the lightweight packaging fraction deposited erroneously in the mixed container was reduced.

Thus, a smaller amount of waste from the mixed fraction leaves the municipality, which supposes the reduction of the costs of transport and the costs of treatment, as well as the reduction of the annoyances caused to the neighbors of the municipality by odors coming from the mixed container and of the neighboring municipalities by the passage of trucks loaded with organic wastes in phase of decomposition. When fewer tonnes of wastes are delivered to incinerate and more tonnes of waste separated correctly, there is a reduction in the total cost of the collection service [32].

It should be taken into account that the improvement of recovery and recycling data corresponds to the implementation of the decentralized model in which it is estimated to participate in around 20% of the population of the municipality. Therefore, the participation of a greater

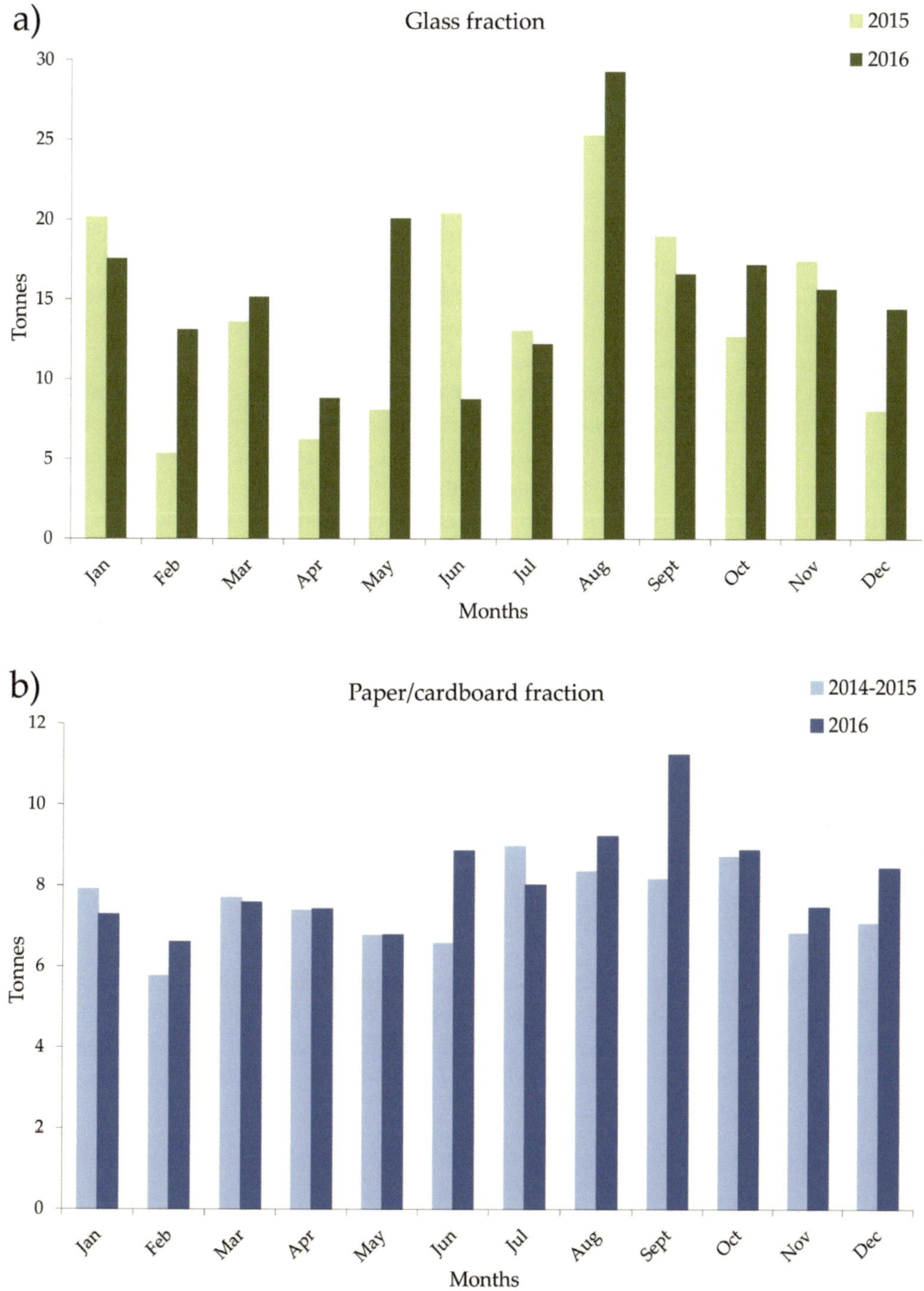

Figure 9. Different fractions collected in the municipality of Allariz: (a) monthly production of the year 2015 vs. monthly production of the year 2016 of the glass fraction, (b) average monthly production of the years 2014–2015 vs monthly production of the year 2016 of the paper-cardboard fraction.

number of inhabitants in the different systems of composting will enable the improvement of the recycling data.

3. Conclusions

Through the promotion of decentralized composting, carried out in the municipality of Allariz, the organic fraction of municipal waste is valued by obtaining high-quality compost and closes the circle of organic matter by applying it to the soil. This, in addition to increasing the environmental awareness of citizens, achieves the diversion of biowaste from the energy recovery, reducing the emissions of greenhouse gases emitted by the sector during the incineration and the landfill. In this way, a greater waste exploitation is achieved in accordance with the hierarchy imposed by European regulations, consolidating an efficient waste management model capable of contributing to sustainable development. On the other hand, the saving of economic resources is reinvested in the locality and it is possible to strengthen the employment in a rural region and to minimize the dependence of the services provided by supramunicipal organisms.

Currently, in Galicia, it is very common to burn pruning and gardening remains produced in the houses. By means of composting, a recovery of these wastes is achieved due to the need to add a low-density material, with capacity to retain water and to contribute to the porosity, to the mixture of household biowaste.

Another indirect result of the establishment of decentralized biowaste composting is the increase in the percentage of separate collection of the other municipal waste fractions. It is because, thanks to direct contact with the citizens, channels of information and learning are established, which improve the separation of the different fractions of both organic and inorganic wastes. Therefore, the reduction in the mixed fraction is not only a result of the removal of the biowaste therefrom but also of the improvement in the separation of the other fractions.

Due to its presence in our society, the management of organic matter, both biowaste and green waste, is the cornerstone of good municipal waste management. This makes European legislation more and more demanding and aims to achieve more rigorous objectives. The economy of the future depends on the degree of sustainability applied to the management of these vital resources.

Acknowledgements

This study was financially supported by the local government of Allariz. The authors thank the research support services of the University of Vigo (CACTI) for the carbon, nitrogen and heavy metals analysis. The authors also thank Xosé Romero, environment technician of Allariz, for his work.

Author details

Iria Villar Comesaña[1]*, David Alves[1], Salustiano Mato[1], Xosé Manuel Romero[2] and Bernardo Varela[2]

*Address all correspondence to: iriavillar@uvigo.es

1 Department of Ecology and Animal Biology, University of Vigo, Vigo, Spain

2 Local Government of Allariz, Spain

References

[1] Directive 2008/98/EC of the European Parliament and of the Council of 19 November 2008 on Waste and Repealing Certain Directives. The Official Journal of the European Union Law. 312, p. 3-30. Available from: http://eur-lex.europa.eu/legal-content/ES/TXT/?uri=celex%3A32008L0098

[2] Belton V, Crowe DV, Matthews R, Scott S. A survey of public attitudes to recycling in Glasgow (UK). Waste Management and Research. 1994;**12**:351-367

[3] Dahlén L, Vukicevic S, Meijer J-E, Lagerkvist A. Comparison of different collection systems for sorted household waste in Sweden. Waste Management. 2007;**27**:1298-305

[4] Purcell M, Magette W. Attitudes and behaviour towards waste management in the Dublin, Ireland region. Waste Management. 2010;**30**:1997-2006

[5] Supriyadi S, Kriwoken LK, Birley I. Solid waste management solutions for Semarang, Indonesia. Waste Management and Research. 2000;**18**:557-566

[6] Henry RK, Yongsheng Z, Jun D. Municipal solid waste management challenges in developing countries-Kenyan case study. Waste Management. 2006;**26**:92-100

[7] Oberlin AS, Szántó GL. Community level composting in a developing country: Case study of KIWODET, Tanzania. Waste Management and Research. 2011;**29**:1071-1077

[8] Jofra M, Citlalic A, Calaf M. Estudio sobre modelos de gestión de residuos en entornos rurales aislados. ENT Environment and Management. 2011. Available from: http://www.mapama.gob.es/es/calidad-y-evaluacion-ambiental/publicaciones/ruralesaislados.aspx

[9] Ley 22/2011, de 28 de julio, de residuos y suelos contaminados. Boletín Oficial del Estado 2011; 181, p. 85650-85705. Available from: https://www.boe.es/buscar/doc.php?id=BOE-A-2013-7540

[10] Gallardo A, Bovea MD, Colomer FJ, Prades M, Carlos M. Comparison of different collection systems for sorted household waste in Spain. Waste Management. 2010;**30**:2430-2439

[11] Xunta de Galicia. Plan de Gestión de Residuos Urbanos de Galicia 2010-2020 (PGRUG) [Internet]. 2011. Available from: http://sirga.xunta.gal/c/document_library/get_file?folderId=190428&name=DLFE-16056.pdf

[12] Zurbrügg C, Drescher S, Patel A, Sharatchandra H. Decentralised composting of urban waste – An overview of community and private initiatives in Indian cities. Waste Management. 2004;**24**:655-662

[13] Deputación de Pontevedra. "Revitaliza" Program [Internet]. 2017. Available from: http://www.depo.es/es/plan-compost-revitaliza/que-e-o-compost

[14] Zucconi F, De Bertoldi M. Compost specifications for the production and characterization of compost from municipal solid waste. In: De Bertoldi M, Ferranti MP, L'Hermite P, Zucconi F, editors. Compost: Production, Quality and Use. London: Elsevier Applied Science Publisher; 1987. pp. 30-50

[15] Casco JM, Herrero RM. Compostaje. Madrid: Ediciones Mundi-Prensa; 2008

[16] Maulini-Duran C, Artola A, Font X, Sánchez A. Gaseous emissions in municipal wastes composting: Effect of the bulking agent. Bioresource Technology. 2014;**172**:260-268

[17] Ansorena J. El compost de biorresiduos. Normativa, calidad y aplicaciones. Madrid: Ediciones Mundi-Prensa; 2016

[18] Mato S, Otero D, Garcia M. Composting of <100 mm fraction of municipal solid waste. Waste Management and Research. 1994;**12**:315-325

[19] González-Torre PL, Adenso-Díaz B. Influence of distance on the motivation and frequency of household recycling. Waste Management. 2005;**25**:15-23

[20] Storino F, Arizmendiarrieta JS, Irigoyen I, Muro J, Aparicio-Tejo PM. Meat waste as feedstock for home composting: Effects on the process and quality of compost. Waste Management. 2016;**56**:53-62

[21] Kalamdhad AS, Singh YK, Ali M, Khwairakpam M, Kazmi AA. Rotary drum composting of vegetable waste and tree leaves. Bioresource Technology. 2009;**100**:6442-6450

[22] Bonhotal J, Schwarz M, Feinland G. In-vessel composting for medium-scale food waste generators. BioCycle. 2011;**52**:49-53

[23] Real Decreto 506/2013, de 28 de junio, sobre productos fertilizantes. Boletín Oficial del Estado 2013; 164, Sect. 1, p. 51119-51207. Available from: https://www.boe.es/buscar/doc.php?id=BOE-A-2013-7540

[24] Thompson WH, Leege PB, Millner PD, Watson ME, editors. TMECC. Test Methods for the Examination of Composting and Compost. Bethesda, MD: Composting Council Research and Education Foundation, and US Department of Agriculture; 2002

[25] Bernal MP, Paredes C, Sánchez-Monedero MA, Cegarra J. Maturity and stability parameters of composts prepared with a wide range of organic wastes. Bioresource Technology. 1998;**63**:91-99

[26] Zucconi F, Monaco A, Forte M, Bertoldi M. Phytotoxins during the stabilization of organic matter. In: Gasser JK, editor. Composting of Agricultural and Other Wastes. London: Elsevier Applied Science Publisher; 1985. pp. 73-85

[27] Zucconi F, Pera A, Forte M, de Bertoldi M. Evaluating toxicity of immature compost. BioCycle. 1981;**22**:54-57

[28] Moldes A, Cendón Y, Barral MT. Evaluation of municipal solid waste compost as a plant growing media component, by applying mixture design. Bioresource Technology. 2007;**98**:3069-3075

[29] Villar I, Alves D, Mato S. Seafood-processing sludge composting: Changes to microbial communities and physico-chemical parameters of static treatment versus for turning during the maturation stage. Plos One. 2016;**11**:e0168590–e0168590

[30] Haug RT. The Practical Handbook of Compost Engineering. Boca Raton: Lewis Publishers; 1993

[31] Adhikari BK, Barrington S, Martinez J, King S. Characterization of food waste and bulking agents for composting. Waste Management. 2008;**28**:795-804

[32] Gomes AP, Matos MA, Carvalho IC. Separate collection of the biodegradable fraction of MSW: An economic assessment. Waste Management. 2008;**28**:1711-1719

Chapter 5

Solid Waste in Agricultural Soils: An Approach Based on Environmental Principles, Human Health, and Food Security

Cácio Luiz Boechat,
Adriana Miranda de Santana Arauco,
Rose Maria Duda,
Antonny Francisco Sampaio de Sena,
Manoel Emiliano Lopes de Souza and
Ana Clecia Campos Brito

Additional information is available at the end of the chapter

http://dx.doi.org/10.5772/intechopen.69701

Abstract

In recent decades, projections involving population growth, changes in consumption patterns, modifications of the wastes produced, and a significant increase in resource extraction have caused concern in the scientific world, in treatment companies, and in environmental and governmental agencies throughout the world, regarding the destination of the large volume of solid wastes generated, the relatively high contents of potentially toxic, carcinogenic and mutagenic substances and pathogenic microorganisms. Waste management has become very important to ensure elementary resources such as water, phosphorus, and food in the future. The recycling policy thus requires that wastes be classified in terms of their appropriateness for new uses and also based on their origins and hazardousness of handling. These classifications are essential in order to allow a minimum of rationality in their new destinations. Currently, several studies have been performed to use solid wastes from human activities as soil conditioners and/or fertilizers for increasing crop productivity. Therefore, studies that monitor organic waste effects on agricultural soils deserve the attention of the international scientific community, as it enables increases in the productivity of agricultural crops, fiber, and biomass energy combined to reduce risks to human, plant, and animal health and environment.

Keywords: organic fertilizer, human health, environmental security, agricultural approach

1. Introduction

The term "organic waste" commonly used in the literature refers to any type of solid, semi-solid, gaseous, or liquid organic materials with different physicochemical characteristics, with a highly complex composition. In this chapter, we shall discuss the term regarding the possible reuse and use of organic solid wastes (SWs) in agriculture and the challenges of this proposal.

In Brazil, as in other countries in South America, Latin America, North America, and Europe, the combination of industrial development, demographic pressure, and increased consumption by the population has caused significant increments in the volume of municipal solid wastes. In the context of the change in consumption patterns and urban and industrial development, sustainable management of waste is one of the most important aspects of planning urban infrastructure, since, without sustainable waste management there would be risks to the environment, human health, quality of life, and the economy.

Water, which up to the last generation was considered an abundant natural resource, has become a limiting factor that was compromised because of high pollution in some regions, as a result of the inadequate discharge of urban sewage which is today the main polluter of water sources. However, sewage treatment generates a sludge rich in organic matter and nutrients whose final disposal in the environment should be planned already during the planning phase of Treatment Plants, thus avoiding partially canceling out the benefit of effluent collection and treatment [1].

The evaluation of processes that protect the environment alongside resource and energy consumption from the most favorable to the least favorable actions is known as waste management hierarchy. The waste hierarchy is a set of priorities for the efficient use of resources, such avoidance including action to reduce the amount of waste generated by households, industry, and all levels of government, resource recovery including reuse, recycling, reprocessing, and energy recovery, consistent with the most efficient use of the recovered resources and disposal including management of all disposal options in the most environmentally responsible manner. However, the waste hierarchy recognizes that some types of waste, such as hazardous chemicals or asbestos, cannot be safely recycled and direct treatment or disposal is the most appropriate management option [2].

Conceptually, it is essential to consider the disposal, not the discarding of wastes. The former, disposal involves an organized action for the purpose of using and not only eliminating wastes, and reutilization of waste is definitely the most useful option from the economic, environmental, and social standpoint. The second, discarding, on the other hand, is defined as the act or effect of getting rid of something that is no longer useful or which is no longer wanted, or even anything that that is separated because it has been rejected or set aside. Thus, discarding is performed randomly, without great care, and the main interest is to get rid of the waste [3].

Currently, worldwide, organic wastes are most commonly disposed of in controlled landfills, incineration, and applying them to agricultural soils (e.g., home composting, central composting plants). Since incineration is a very expensive and environmentally criticized technique, other recycling or reusing options are considered better.

It is estimated that currently there are about 3 billion inhabitants generating 1.2 kg per person per day, almost 1.3 billion tons of municipal solid waste (MSW) a year or 1.2 kg per capita per day. In 2025, this will probably increase to 4.3 billion tons from urban residents, about 1.42 kg per inhabitant per day of MSW (2.2 thousand millions of tons a year). However, they are highly variable since there are differences in the rates of waste generation between countries, between cities, and even within cities [4].

In Brazil, with a population of 206 million inhabitants, daily about 218,874 thousand tons of urban solid wastes (USWs) are produced, generated in the country, and the main form of final disposal of USW is in sanitary landfills (58.7%). According to IBGE, rural occupation in Brazil corresponds to 30 million people, which is approximately 15% of the total population of the country. And rural garbage collection is insufficient, since it only covers 20% of the domiciles in the country. In general, a rural waste collection system is inefficient, and the wastes are discarded in the environment, burned in most cases, or simply dumped in the open, due to the lack of waste collection and treatment.

Many countries around the world have been incorporating the organic wastes from sewage treatment plants (STPs) or from the selective collection of urban garbage into the soil for several decades now, and have created and altered a few preventive rules against possible problems with the contaminants present in them, emphasizing potentially toxic metals, organic contaminants, and pathogens.

2. Definition and classification of wastes

The intensification of the industrial process and the rapid population growth and the consequent demand for consumer goods have provoked an increase in the volume of wastes generated. Therefore, there is a concern, on a global scale, to solve the problem of excessive generation and environmentally safe final disposal.

Considering the complexity involved in the discussion of the concept of solid wastes, it is important to begin by performing a comparative analysis of the terms: garbage and waste. Garbage is a polysemic term which is related to several words and can be interpreted in different manners, varying in time and in space according to the socioeconomic and cultural contexts in which it is used. From the semantic standpoint, it corresponds to all of the useless materials, all materials discarded in a public place, everything that is "thrown away," in other words, old objects without value [2].

The National Policy of Solid Wastes (NPSW, Brazil), Law 12,305/10, presents a broad definition of solid wastes, including gases and liquids, as described in paragraph XVI of article 3.

> *[...] discarded material, substance, object or goods resulting from human activities in society, whose final disposal is done, proposed to do or mandatorily performed in the solid or semi-solid states, as well as gases in containers and liquids whose specificities make in impossible to discharge them into the public sewage system or into bodies of water, or that for this require solutions that are technical or economically unfeasible considering the best technology available [2].*

Garbage and waste have different connotations and may be understood as byproducts generated by the different human activities. The difference lies in the relationship between people and the material to be discarded, since although garbage can be reused, people consider it something useless and valueless that must be thrown away. On the other hand, waste (residue) is seen as material with commercial value that can be reutilized to produce new products [5, 6].

The topic of "wastes" has been a priority since the Second United Nations Conference on Environment and Development (UNCED), which also became known as ECO-92 or Rio-92, because it took place in the city of Rio de Janeiro in 1992. It was a conference on a global scale, both of the rich countries and the poorer ones, because it contributed directly or indirectly to global warming and climate changes [7]. At this conference, Agenda 21 was elaborated, a document that includes among its programs a few actions relating to the management of urban-industrial solid wastes. In this document, the management of domestic solid wastes must include not only its disposal or even its reuse, but also the adoption of measures that will be able to alter society's patterns of production and consumption. Furthermore, every country and city must establish programs to comply successfully with the agreement, according to local conditions and even their economic capacity [8].

However, the management of solid wastes in urban areas is based historically on the linear logic that considers collection as the removal of wastes from the vicinity of the population and final disposal as putting them on soil in garbage dumps and landfills. This concept, besides manifest pollution in all forms, has led to the saturation of sites for the final disposal of the wastes [6, 9].

According to Ref. [10], in society, there is a lack of social concern about "rural garbage" with a mistaken notion of the urban population over the rural one, in which the former considers that due to the small number of people who live in the countryside—approximately 19% of the population—garbage is an insignificant problem. However, one does not perceive that this environmental damage in the rural area has major consequences on the quality of life of the urban zones, including water supplied to the cities. These types of wastes are generated by various activities, but if they are not well managed, they may cause various types of environmental damage [11].

Besides the significant increase in the generation of these wastes, in recent years there have been significant changes in their composition and characteristics and increased hazardousness [12]. These changes are the result especially of the development models with programmed obsolescence and discardability of the products and the change in consumption patterns based on excessive and superfluous consumption.

There are several ways of classifying the solid wastes. The recycling policy thus requires that wastes be classified in terms of their appropriateness for new uses, and also based on their origins and hazardousness of handling. These classifications are essential in order to allow a minimum of rationality in their new destinations. As every country has its specific classification, in Brazil the Brazilian Association of Technical Standards [13], through NBR n° 10.004/2004, establishes that solid wastes can be classified regarding their hazardousness as: Class I, hazardous; Class II, noninert; Class III, inert. As follows:

http://dx.doi.org/10.5772/intechopen.69701

Class I or hazardous: Those that because of their intrinsic characteristics of flammability, corrosiveness, toxicity, or pathogenicity present risks to public health due to the increased mortality or morbidity, or else have adverse effects on the environment when handled or disposed of inadequately.

Class II or noninert: These are wastes that may present characteristics of combustibility, biodegradability, or solubility, which may result in risks to health or to the environment, and which cannot be included in the classifications of Class I wastes, hazardous or Class III, inert.

Class III or inert: Due to their intrinsic characteristics, these do not offer a risk to health and to the environment and when sampled representatively, according to Standard NBR 10,006 and submitted to a static or dynamic contact with distilled or deionized water, at ambient temperature, with a solubilization test according to Standard NBR 10,006, none of their constituents are solubilized at concentrations higher than the water potability standards, according to List number 8 (Annex H of NBR 10.004), except for the standards of aspect, color, turbidity, and taste.

Another criterion for classification looks at the origin of the wastes, i.e., the generating sources. According to the Manual for Integrated Management of Solid Wastes [14], wastes can be classified, as to generating source, into three categories: urban solid wastes, industrial solid wastes (ISWs), and special wastes.

Urban solid wastes (USWs) are wastes resulting from households (domiciliary or domestic), health service wastes, civil construction wastes, wastes from pruning and grass-cutting, wastes from ports, airports, bus terminals, train terminals, and the wastes of services that cover the commercial wastes, wastes from cleaning manholes, and wastes from sweeping, markets, and others [15].

Industrial solid wastes (ISWs) include the wastes of processing industries, radioactive wastes, and agricultural wastes. They are extremely varied and present diversified characteristics, since they depend on the kind of product manufactured and must therefore be studied case by case. Radioactive wastes (nuclear wastes) are those that emit radiation above the limits allowed by Brazilian standards, generally resulting from nuclear fuels, which, according to the legislation that specifies them, are in the exclusive purview of the National Committee of Nuclear Energy.

Agricultural wastes are those generated by activities pertaining to agriculture and livestock, such as containers of fertilizers, agricultural pesticides, feed, remnants of harvests, and manure. Since agrochemical containers are highly toxic they have specific legislation.

Some wastes are also considered special because of their differentiated characteristics, which include tires, batteries, and fluorescent lamps.

Generally, urban solid waste in Brazil is composed of 61% of organic matter, 15% of paper, 15% of plastic, 3% of glass, 2% of metal, and 5% of others. Despite meeting the specific legislation of each municipality, commercial garbage, up to 50 kg or liters, and domiciliary garbage are the responsibility of the city administrations, while the others are the responsibility of the generator himself. The wastes generated in rural, industrial, and residential activities, such as packaging and batteries, products that no longer work, and others, are the responsibility of the company that manufactured them, and this company must collect and dispose correctly of this material [2].

When solid wastes (SWs) are badly managed, they become a sanitary, environmental and social problem. The basic instrument to manage them is to know the sources and types of solid wastes through data on their composition and rate of generation [16]. However, the composition and rate of generation of solid wastes are a function of a number of variables, including the socioeconomic situation of the population, the degree of industrialization of the region, its geographical location, the sources of energy, and the climate [17].

The law of the National Policy on Solid Wastes [2] defines as its objective to establish regulations for the disposal of wastes, the responsibility of the manufacturers, of the consumers, and of the authorities. As regard the agricultural sector, the law establishes that the reverse logistics system should be applied. This is a tool for economic and social development characterized by a set of actions, procedures, and means destined to make it feasible to collect and restitute solid wastes to the business sector, for reuse, in its cycle or in other production cycles, or another environmentally appropriate final disposal. In the rural area, this instrument is applied to pesticides, its wastes and packaging, as well as to other products whose packaging, after use, constitute dangerous wastes. There may be shared management of the urban and rural wastes, involving the manufacturers, importers, distributors, and vendors, consumers, and heads of the public cleaning services.

3. Waste management and use in agricultural soils

Worldwide daily millions of tons of solid wastes are generated, which must be collected, selected, treated, and disposed of appropriately. In China, India, and other countries, such as Turkey, Mexico, and Brazil, almost 90% of the solid wastes that are composed mainly of the organic fractions are usually sent to landfills and garbage dumps, freely releasing huge amounts of CO_2 and CH_4 into the atmosphere [18, 19].

Waste management in urban and rural areas is one of the great challenges to Public Administration and Society. The National Policy on Solid Wastes (NPSW, Law 12,305/2010) [2] encouraged considerable changes in solid waste management in Brazil. According to Ref. [20], among the various challenges for NPSW is sending wastes mandatorily to recycling and composting of the organic fraction of the urban solid wastes (USWs). The organic fraction of USW should not be placed in the landfill but improved by biological treatment [21]. And composting appears as one of the most promising alternatives for an essentially agricultural country like Brazil, and is very important because it allows the recycling of the organic molecules that have a nutritional function and also because it diminishes the polluting and contaminant potential of the wastes [22].

Domestic solid wastes in Brazil present a high percentage of organic residues formed by remnants of food, and fruit and vegetable peels and even gardening wastes, but composting of organic wastes present in the urban garbage is relatively rarely practiced [22]. According to Ref. [23] cited by Ref. [21], in Brazil there are 211 composting plants in operation. They receive urban, industrial, agricultural, and forestry wastes. Each of these plants has a capacity to recycle an average of 10,000 tons a year, but this amount is too small to cover the total need for the treatment of wastes generated in Brazil.

Sewage sludge is a material that results from the primary and secondary sewage treatment processes and it has a highly complex composition. Because of its composition, which is rich in organic matters, nitrogen and phosphorus, sewage sludge has been strongly suggested for use in agriculture as a soil conditioner and fertilizer. The benefits that could be obtained from

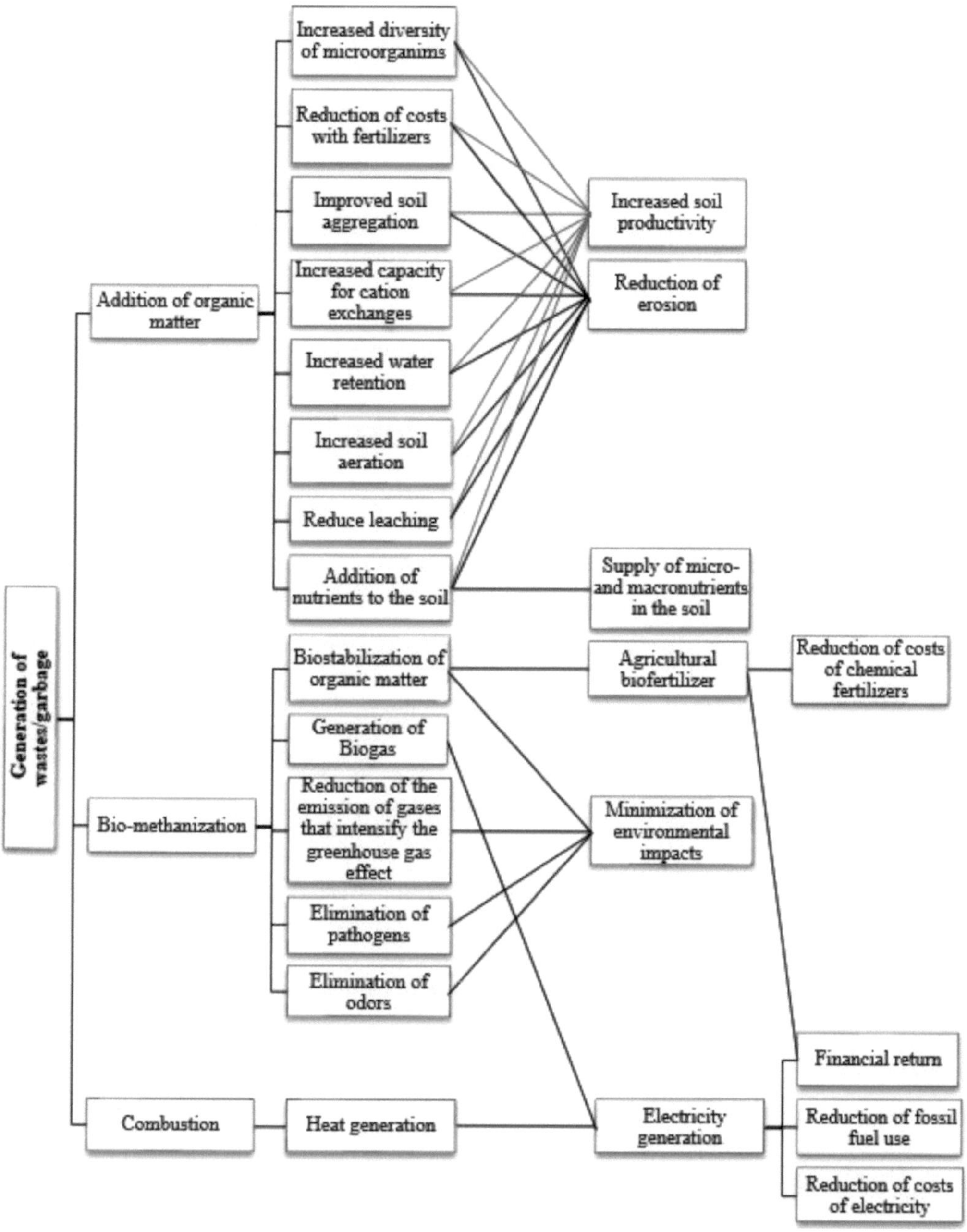

Figure 1. Positive impacts of using wastes (adapted from Ref. [26]).

its use would be the recycling of organic matter and supplying nutrients to soil, improving its physical, chemical, and biological properties and agricultural productivity. However, since the sludge contains high concentrations of contaminants, this practice may result in the direct addition of various pathogens and undesired chemical substances to agricultural soil and consequently to the food chain [24].

In Brazil, the criteria and procedures for the agricultural use of sewage sludge generated in sanitary sewage treatment plants and its byproducts are defined in National Council for the Environment (CONAMA) Resolution 375/2006 [25]. Among the criteria for sewage sludge use, Article 12 establishes that

> *"It is forbidden to use any class of sewage sludge or byproduct for pastures and the cultivation of vegetables, tubercles, roots and flooded cultures, as well as the other cultures whose edible part is in contact with the soil."*

Recycling wastes in agricultural soils is emphasized. This is a much-used alternative in several countries, such as the United States, Holland, Australia, and others. The use of wastes in agriculture (applying them on soils in a controlled manner) and the generation of energy, for instance, may mean environmental and economic gains, besides minimizing the negative impacts of the disposal and inadequate discharge, as described in **Figure 1**.

For instance, the use of urban organic wastes is disseminated worldwide as fertilizers and/or soil conditioners. Garbage compost and sewage sludge are outstanding among them. It is also worthwhile mentioning among the organic wastes those from agribusiness, because, due to their origin, there is a low probability of contaminants in their composition. A good example for this class is wastes generated by the sugar and alcohol industry, filter cake, soot, and vinasse, which are recycled in the agricultural areas of the plant itself [3].

Filter cake presents a high percentage of humidity (70–80%) and high contents of organic matter and phosphorus, besides nitrogen, calcium, and potassium, and it is used mainly in sugar cane plant fertilization with savings for the farmer in the costs of implementing this crop [27].

4. Environmental risk and public health

The use of municipal waste is a source of fertilizers and correctives in agriculture and it is one of the alternatives used worldwide to minimize this conflict. However, long-term studies are needed to evaluate the potential and impact of using wastes from different human activities on the quality of agriculture, environment, and human health.

Wastes as a source of plant nutrients have been used in some countries for decades [28–30]. However, research has shown that applying these wastes to the soil may cause environmental problems due to the addition of excess N, pathogens, heavy metals [31–34], acidification [35], and salinization of agricultural soils [36].

Among the various substances that contaminate water and soil, heavy metals have aroused concern among the population because of their high toxic, mutagenic, and carcinogenic power [37]. Heavy metals are elements with an atomic number greater than 20, and a density greater than 5 mg cm^{-3}. This group includes any element that could damage the plant and animal organisms, including metalloids and semimetals such as selenium (Se) and arsenic (As) [38]. All these elements have high reactivity and under normal conditions are traces in the mineralogical composition of soils [39]. Therefore, their initial concentrations in soil depend on the composition of the bedrock and on the pedogenetic processes that originated it [40].

Heavy metals have important functions in the biosphere, acting as essential micronutrients for plants (Cu, Fe, Mg, and Zn), or as beneficial (Mo and possibly Ni) to microorganisms (Co and Mo genus Rhizobium bacteria) and to animals (Co, Cr, Mo, Cu, Se, and Zn) [41, 42]. However, when these elements are found at high concentrations they are toxic to superior organisms, just like nonnutritional elements or without biological effect, such as Cd, Pb, and As [41].

All heavy metals are toxic to the biological systems. The level of risk is a function of the quantity of contaminant and the time the organism is exposed to it [43]. **Table 1** shows the heavy metal ions, the main sources of pollution and the toxicological characteristics caused by exposure to As, Cd, Cr, Cu, Zn, and Pb.

Domestic sewage sludge and biosolids that are generated in cities due to household garbage may present over 140 types of enteric viruses that cause diseases, especially in children. Pathogenic

Ions	Main sources of pollution	Toxicological characteristics
Arsenic (As)	Herbicides, insecticides, fungicides, mining, and glassware. Paints and dyes industry.	Generally, the inorganic compounds are considered more toxic. As^{3+} is more toxic than As^{5+} at least at very high doses. Various organs and tissues are affected, such as the skin, respiratory system, cardiovascular system, reproductive system, gastrointestinal system, and nervous system [44].
Cadmium (Cd)	Industrial effluents, electroplating, production of pigments, electronic equipment, lubricants, photographic accessories, insecticides, and fossil fuels.	In rat studies in which the respiratory tract of the animals was exposed continuously to an aerosol with a low concentration of $CdCl_2$, a high incidence of lung cancer was observed and evidence was shown of the relationship between dose and response. High levels of Cd inhalation cause lethal pulmonary edema [45].
Chromium (Cr)	Industrial effluents, production of aluminum and steel, paints, pigments, explosives, paper, and photography.	The toxic effects of Cr^{3+} occur only through parenteral administration. Humans and other animals, when exposed to Cr, develop cancer. Cr^{6+} in the diet affects the gastrointestinal tract, the kidneys, and the hematological system and causes several genetic damages. In some studies, the $CrCl_3$ was found accumulated in the cell nucleus [46, 47].
Copper (Cu)	Pipe corrosion, domestic sewage, algicides, fungicides, pesticides, mining, foundries, and metal refinement.	Few cases of acute effects of Cu have been reported. Among them the main ones are gastric burning, nausea, vomiting, diarrheas, lesions of the gastrointestinal tract, and hemolytic anemia. Chronic effects are rarely reported, except for Wilson's disease, responsible for the accumulation of copper in the liver, brain, and kidney [48].

Ions	Main sources of pollution	Toxicological characteristics
Zinc (Zn)	Electroplating, mining, wood combustion, waste incineration, domestic sewage and iron, and steel production.	The accumulation of Zn does not cause profound deficiencies. For this reason, it is considered as having low toxicity. The excessive intake of Zn may provoke gastrointestinal disorders and diarrhea [49].
Lead (Pb)	Industrial effluents, tobacco, paints, pipes, metallurgy, and electrodeposition industry.	After the Pb is absorbed by human body it can be found in the blood, soft, and mineralized tissues [50]. According to Ref. [51], for neurological, metabolic, and behavioral reasons children are more vulnerable to the effect of Pb than adults. Among their main effects are diminished intelligence quotient, effects on the nervous system, reduction of sensory functions, involuntary nervous and kidney functions, and premature births [52–54].

Table 1. Characteristics intrinsic to the ions of heavy metals and risk to human health.

organisms that cause infections depend on the resistance of organisms to sewage treatment, and the environmental conditions, the dose of infection, pathogenicity, susceptibility, degree of host immunity, and degree of human exposure to the foci of transmission [55].

The use of sewage sludge presents several good results. However, in some cases, the sludge may be harmful to the plants and also worsen the diseases due to the presence of pathogenic microorganisms, mostly saprophytes. The main pathogens present in the sludge are bacteria, viruses, and parasites (**Table 2**). The quantity of these microorganisms is very variable, depending on the time and season of the year. In order to use the sludge in agriculture, it is necessary to characterize and quantify the chemical contaminants and pathogenic microorganisms present [56, 57].

Organisms	Disease and symptoms
Bacteria	
Salmonella sp.	Salmonellosis (of the typhoid fever type)
Shigella sp.	Bacillary dysentery
Vibrio cholerae	Cholera
Campylobacter jejuni	Gastroenteritis
Escherichia coli patogênica	Gastroenteritis
Enteric viruses	
Hepatitis A	Infectious hepatitis
Norwalk and Norwalk-like	Gastroenteritis with severe diarrhea
Rotavirus	Acute gastroenteritis

Organisms	Disease and symptoms
Enterovirus	
Poliovirus	Poliomyelitis
Coxsackievirus	Meningitis, pneumonia, hepatitis and fever
Reovirus	Respiratory infection, gastroenteritis
Astrovirus	Gastroenteritis
Calicivirus	Gastroenteritis
Protozoa	
Cryptosporidium	Gastroenteritis
Entomoeba histolytica	Gastroenteritis
Giardia lamblia	Giardia (including diarrhea, abdominal pain and weight loss)
Balantidium coli	Diarrhea
Toxoplasma gondii	Toxoplasmosis
Helminths	
Ascaris lumbricoides	Digestive problems and nutritional disorders, abdominal pain
Trichuris trichiura	Abdominal pain, diarrhea, anemia, weight loss
Toxocara canis	Abdominal discomfort, muscle pains, neurological symptoms
Taenia saginata	Nervousness, anorexia, abdominal pain, digestive disorders
Necator americanus	Ancylostomiasis

Table 2. Organisms that may be present in sewage sludge and are a risk to human health (adapted from Ref. [25]).

Thus, systematic treatment of the sewage sludges or urban wastes before their use in agricultural soils diminishes the risk to human and animal health through infection because it reduces the chances of survival of these pathogenic organisms.

5. Monitoring soil and water

The organic substances present in wastes can maintain or even raise the organic matter content in the soil, which is extremely advantageous from the agronomic standpoint, because the organic fraction is directly connected to a number of functions that are important to maintain soil fertility and quality. However, some wastes, especially those from agriculture and livestock activities, can have in their composition fecal coliforms, a few pathogenic microorganisms (e.g., Salmonella spp) and persistent organic molecules that can cause biological imbalances in soil and water [3]. These substances, however, may originate in other sources, from sanitary inspections of animals, and also from activities such as deworming and the application of agricultural pesticides to crops and pastures [58].

Inorganic substances, on the other hand, are represented mainly by elements that are essential or beneficial to plants and are usually found in wastes from industrial activities, such as mining and steel foundries and from the sanitary treatment sector in urban centers [59]. Some wastes generated in these activities usually present high micronutrient contents, such as the following elements: copper, iron, manganese, and zinc. However, these elements present constantly in unbalanced proportions for plant nutrition. This may promote the practice of high doses, overloading the natural functions of soil and causing imbalances [3].

Thus, wastes with a potential for use in agriculture also present as possible sources of contaminants and are a risk for the quality of soil and groundwater. Therefore, this practice should be verified by the environmental control and monitoring agencies of each state or territory. The most common and effective tool developed by these countries is the formulation of specific laws to monitor and protect the quality of soils and groundwater, which are in turn based on surveys of the critical contents of these substances, as well as studies of their respective potential for damage to the environment and to human health [60, 61].

In Brazil, this monitoring is performed based on the experiences and models practiced in countries like Holland and Germany. The National Council of the Environment (CONAMA) through Resolution n° 420 of December 28, 2009 determined the criteria for the elaboration of guiding values of soil and groundwater quality regarding the presence of chemical substances and established the guidelines for environmental management of areas contaminated by these substances as a result of anthropic activities, and also stipulated the maximum values for the same substances in groundwater [62].

Therefore, according to CONAMA, the guiding values are concentrations of chemical substances that provide guidance regarding the quality and changes in soil and groundwater. The resolution also determines that the guiding values are classified into three groups according to the contents of the elements investigated, which are quality reference value (QRV), prevention value (PV), and investigation value (IV).

The quality reference value or QRV corresponds to the concentration of a given substance that defines the natural quality of the soil, and it is determined based on the statistical interpretation of physicochemical analyses of samples of various types of soil, being used as a reference in actions to prevent soil and ground water pollution and to control contaminated areas.

Prevention value (PV) is the concentration of the limit value of a given substance in the soil, such that it can sustain its main functions. It is predetermined by CONAMA and is used to educate through educational measures and penalties applied to those responsible for possible alterations in the environment.

Investigation value (IV) is the concentration of a given substance in soil or in groundwater, above which there are direct or indirect potential risks to human health, considering a standardized scenario of exposure. For the soil, it is calculated using a procedure to evaluate the risk to human health in different contexts: agricultural, residential, and industrial. When the intervention values are surpassed, immediate actions must be taken due to the finding of a potential risk of a deleterious effect on human health [63].

The QRVs are determined from the survey of natural contents of the elements in soil. For this it is necessary to sample the soil taking into account the diversity of soil classes and the original materials existing in the region, seeking to perform the collections in minimally preserved soils, with little or no apparent signs of anthropic interventions, since the pedogenetic processes and the geochemical formation of each region interfere directly with the natural contents of these substances [42, 64, 65].

Therefore, Brazilian law determines that the QRVs for soils should be established by each state of the federation and it is not recommended to use the values of one state for another state [63]. **Table 3** shows some inorganic and organic substances and their respective guiding values for soils and groundwater in the state of São Paulo, as an example of the model adopted by the country to organize its monitoring tool.

Given the complexity of the relations that may occur between the substances present in the wastes and the soil attributes, current legislation cannot foresee all the long-term scenarios and behaviors of these substances. This is a weakness in the environmental monitoring system that requires constant investment and updating. Ref. [59] emphasize that all components of the agricultural environment should be periodically monitored, such as soil, water, and their biological fractions, in order to avoid problems caused by the intermittent use of wastes for long periods.

Substance	**Soil (mg kg^{-1} dry weight)**					**Groundwater ($\mu g\ L^{-1}$)**
	Quality Reference value (QRV)[1]	**Prevention value (PV)[2]**	**Intervention value (IV)[2]**			
			Agricultural	**Residential**	**Industrial**	
Inorganics						
Antimony	<0.5	2	5	10	25	5
Arsenic	3.5	15	35	55	150	10
Barium	75	120	500	1300	7300	700
Boron	-[3]	nd	nd	nd	nd	2400
Cadmium	<0.5	1.3	3.6	14	160	5
Lead	17	72	150	240	4400	10
Copper	35	60	760	2100	10,000	2000
Mercury	0.05	0.5	1.2	0.9	7	1
Molybdenum	<4	5	11	29	180	30
Nickel	13	30	190	480	3800	70

Substance	Soil (mg kg^{-1} dry weight)					Groundwater (μg L^{-1})
	Quality Reference value (QRV)[1]	Prevention value (PV)[2]	Intervention value (IV)[2]			
			Agricultural	Residential	Industrial	
Nitrate	-	nd	nd	nd	nd	10,000
Silver	0.25	2	25	50	100	50
Selenium	0.25	1.2	24	81	640	10
Zinc	60	86	1900	7000	10,000	180
Volatile aromatic hydrocarbons						
Benzene	na	0.002	0.02	0.08	0.2	5
Styrene	na	0.5	50	60	480	20
Ethylbenzene	na	0.03	0.2	0.6	1.4	300
Toluene	na	0.9	5.6	14	80	700
Organochlorinated pesticides						
Aldrin	na	0.02	0.4	0.8	6	0,03
Endrin	na	0.001	0.8	2.5	17	0,6
Carbofuran	na	0.0001	0.3	0.7	3.8	7

[1]Value determined by the State.
[2]Determined by CONAMA.
Note: nd: not determined in the legislation; na: not applicable to organics.

Table 3. Guiding values for soils and groundwater in the state of São Paulo (adapted from Ref. [66]).

It should also be pointed out that the model adopted to establish the legislation in countries with a tropical climate is based on the experiences of developed countries, usually situated in temperate climate regions, which justifies the need to continue studying the behavior of the potentially toxic substances and their relations with the soil, groundwater, and their biological agents under local climatic conditions, in order further improve the monitoring mechanisms for risk activities such as the reuse of wastes in the agricultural and livestock chains of countries with a tropical climate.

6. Potential of productivity for the agricultural crops

Liming is the first process to be performed to prepare the soil for crops. The purpose of the technique is to correct the soil pH, elevating base saturation, and reduction of exchangeable aluminum in the soil solution to levels appropriate to the crops. Currently, several studies have been performed to use solid wastes from human activities as an alternative substitute to correct soils for cultivation. Outstanding among the wastes studied are steel mill slag and urban sewage sludges. These wastes have proved to be an excellent substitute for limestone, since besides correcting the pH they are a major source of nutrients for the plants [67–70].

Refs. [71, 72] using steel mill slag to produce guava seedlings found that besides the corrective action of pH, the rise in the sum values base saturation, the waste proved to be a source of micronutrients such as zinc, copper, manganese, and boron. There was also a positive effect on the concentrations of calcium, magnesium, and phosphorus in the roots and aerial parts of the seedlings. In addition, using this waste is limited when it contains traces of metals, such as lead and chromium in its constitution.

Ref. [73] incorporate domestic sewage sludge and sewage sludge from a dairy establishment into the soil, observed increased macronutrients in the aerial part and roots of Physic Nut Seedlings (*Jatropha Curcas* L.) plants in both treatments. In these soils, they found that using dairy waste sludge raised the pH in soil, while domestic sludge did not change the soil pH. According to the authors, the initial treatment of dairy sewage sludge consists basically in a biological process with the addition of limestone or lime ($CaCO_3$ or CaO) to eliminate pathogens and promote waste stabilization. Therefore, the carbonate reacted with the hydrogen present in the soil solution (liquid phase of the soil) and water and CO_2. Meanwhile, the domestic sewage sludge did not receive this conditioner at the treatment plant.

The next stage of soil preparation is fertilizing, which consists in applying nutrients in forms available for root absorption. According to Ref. [73], the organic matter in plant nutrition may certainly be substituted, with even better yields, by chemical fertilizers, but their indirect effect on crop development cannot be substituted, whatever the chemical industry product. For this reason, organic matter from urban solid wastes, which constitute approximately 90% of the total mass of sanitary landfill wastes, as well as the domestic sewage sludges can and should be used as a source of nutrients in plant production, although it is necessary to have a greater volume than that of the mineral fertilizers because of the lower concentration of nutrients, although greater attention is required as to the presence of organic, mineral toxic contaminants, or pathogenic microorganisms.

If we perform a very simple analysis of the dynamics of nutrients in the production process in agroecosystems, we conclude that, biomass and nutrients are removed in the harvesting, breaking down the efficient cycling that would occur in natural environments. In this way, the external supply of nutrients, be it in a concentrated chemical form or organic, is necessary to maintain this balance. Otherwise, the producer will be performing unsustainable, predatory production, which will impoverish the soil that supports him [74].

Many research studies have demonstrated the rise in crop productivity when they are fertilized with organic wastes [75]. In research using the filter cake to fertilize different lettuce cultivars increased yields were obtained in all cultivars evaluated. In Ref. [76] utilizing sewage and industrial sludges to produce physic nut seedlings they concluded that the wastes used promoted the production of quality seedlings with a significant increment in root length and in the aerial part of the seedlings. Ref. [77] using solid organic wastes (*Copernicia prunifera* waste and chicken litter) and swine wastewater as a source of nutrients for the production of monkfish (*Enterolobium contortsiliquum* (Vell.) Morong) found that organic waste contributes to growth and nutritional balance of the seedlings. The use of these organic wastes as primary source of nutrients consists of an important environmental management practice.

The effects of adding solid wastes to the soil on agricultural production are not limited to supplying nutritional elements. The organic matter it contains acts significantly to improve the physical, chemical, and biological quality of soil, and is a source of energy for the beneficial edaphic fauna, which has an antagonist action on phytopathogens (nematodes, bacteria, and fungi) and for symbionts, such as the diazotrophs (atmospheric N fixing microorganisms), or phosphate solubilizers (mycorrhizal fungi and rhizobacteria).

Worms, insects, and other organisms of edaphic fauna, in the metabolization of this substrate, release cementing substances that help form aggregates in the soil, significantly improving its drainage and aeration and making it easier to store and drain water and to develop the root system of the crops [73].

During the process of decomposing the organic wastes, the cations present in them are released in the soil solution and/or retained on the colloid surfaces, and are available for absorption by the roots. Simultaneously, humus is formed, another product that is very important for the physicochemical improvement of many sandy soils or highly weathered type 1:1 clays that are chemically poor, which commonly occurs in the soils of the Cerrado and Amazon biomes. Therefore, there must be a constant attempt to increase the organic matter in soil, be it through management (minimum cultivation, no till farming with crop rotation, green fertilization, and others) or added in organic fertilization (animal manure, organic compost from sanitary landfills, sewage and tannery sludges, and others).

7. Final considerations

The economic growth rates of various countries keep up with the rapid urban development with the intense production of solid wastes. In this way, public policies should emerge to help adopt a new management logic, which will take into account the sustainability principles of reducing the wastes generated, reuse, and recycling, treatment and environmentally safe final disposal.

There is a broad discussion regarding the appropriate forms of disposal and use of the solid wastes, and their reuse on agricultural soil has been considered the most interesting option, both from the environmental standpoint and the economic one. However, their use in agriculture should be preceded by an analysis of environmental and economic impact and the indiscriminate use of wastes may lead to contamination.

National and international public policies for solid waste management are already a historical advance, taking into account the mandatory regulation of responsibilities by the management, mainly of urban solid wastes; however, this in itself does not guarantee that it will be done, since the consolidation of these policies requires behavioral and cultural changes that will attenuate deep-rooted practices in individual and collective action.

http://dx.doi.org/10.5772/intechopen.69701

Author details

Cácio Luiz Boechat[1*], Adriana Miranda de Santana Arauco[1], Rose Maria Duda[2], Antonny Francisco Sampaio de Sena[1], Manoel Emiliano Lopes de Souza[1] and Ana Clecia Campos Brito[1]

*Address all correspondence to: clboechat@hotmail.com

1 Federal University of Piauí, Brazil

2 Faculty of Technology of Jaboticabal, Brazil

References

[1] Bettiol W, de Camargo OA, Galvão JAH, Ghini R. A disposição de lodo de esgoto em solo agrícola. In: Bettiol W, de Camargo OA. Lodo de esgoto: impactos ambientais na agricultura. Jaguariúna: Embrapa Meio Ambiente; 2006. pp. 17-23

[2] Brasil. Institui a Política Nacional de Resíduos Sólidos; altera a Lei no 9.605, de 12 de fevereiro de 1998 e dá outras providências [Internet]. 2010. Available from: http://www.planalto.gov.br/ccivil_03/_ato2007- 2010/2010/lei/l12305.htm [Accessed: 25-March-2017]

[3] Pires AMM, Mattiazzo ME. Avaliação da viabilidade do uso de resíduos na agricultura. Circular Técnica 19 - Embrapa. 2008;**1516-4683**:1-9

[4] Hoornweg D, Bhada-Tata P. What a Waste: A Global Review of Solid Waste Management [Internet]. 2012. Available from: https://siteresources.worldbank.org/INTURBANDEVELOPMENT/Resources/336387-1334852610766/What_a_Waste2012_Final.pdf [Accessed: 15-February-2017]

[5] Waldman M. Lixo: cenários e desafios. São Paulo: Cortez; 2010. p. 231

[6] Schlindwein JR. O discurso e a prática do gerenciamento de resíduos sólidos urbanos (RSU) em Caxias do Sul/RS [Dissertação]. Brasília: Universidade de Brasília; 2013

[7] Jacobi PR, Besen GR. Gestão de resíduos sólidos em São Paulo: desafios da sustentabilidade. Estudos Avançados. 2011;**25**:135-158

[8] Vieira ACPE, Garcia JR. A gestão de resíduos sólidos domésticos no Brasil a partir da experiência internacional. Revista Economia & Tecnologia. 2012;**8**:57-66

[9] Sáez A, Urdaneta G, Joheni A. Manejo de residuos sólidos en América Latina y el Caribe [Internet]. 2014. Available from: http://www.redalyc.org/articulo [Accessed: March 25, 2017]

[10] Darolt MR. Lixo rural: do problema à solução [Internet]. 2008. Available from: http://www.comcienc ia.br/comciencia/?section=8&edicao=32&id=373 [Accessed: March 07, 2017]

[11] Arunabha M, Fellow E. Gestão de resíduos sólidos rural: questões e ações [Internet]. 2011. Available from: http://www.sanitation.kerala.gov.in/pdf/workshop/Rural_solid_waste_management.pdf [Accessed: February 24, 2017]

[12] OMS – Organização Mundial da Saúde. The World Health Report 2007 – A safer future: Global public health security in the 21st century [Internet]. 2007. Available from: http://www.who.int/whr/2007/en/index.html [Accessed: 03-March-2017]

[13] Associação Brasileira de Normas Técnicas (ABNT). Resíduos Sólidos - Classificação. NBR 10,004. 2a ed. Rio de Janeiro: ABNT; 2004. pp. 71

[14] Monteiro JH. Gestão integrada dos resíduos sólidos: manual de gerenciamento integrado de resíduos sólidos. IBAM. 2001;**200**

[15] Gomes LP. Resíduos sólidos: Estudos de caracterização e tratabilidade de lixiviados de aterros sanitários para as condições brasileiras. Projeto PROSAB. Rio de Janeiro: ABES; 2009. p. 360

[16] IPEA. Pesquisa sobre pagamento por serviços ambientais urbanos para gestão de resíduos sólidos. Relatório de Pesquisa. Brasília: Instituto de Pesquisa Econômica Aplicada - IPEA; 2010. pp. 1-66

[17] Hoornweg D, Bhada-Tata P. What a waste: A global review of solid waste management. Washington: Urban Development & Local Government Unit, World Bank. 2012; p. 98

[18] Potdar A, Singh A, Unnikrishnan S, Naik N, Naik M, Nimkar I, Patil V. Innovation in solid waste management through Clean Development Mechanism in India and other countries. Procedia Environmental Sciences. 2016;**35**:193-200. DOI: 10.1016/j.psep.2015.07.009

[19] Lino FAM, Ismail KAR. Energy and environmental potential of solid waste in Brazil. Energy Policy. 2011;**39**:3496-3502

[20] Siqueira TMO, Assad ML, Ribeiro CL. Compostagem de Resíduos Sólidos Urbanos no Estado de São Paulo. Ambiente & Sociedade. 2014;**18**:153243-153264

[21] Santos ATL, Henrique NS, Shhlindwein JA, Ferreira E, Stachiw R. Aproveitamento da fração orgânica dos resíduos sólidos urbanos para produção de composto orgânico. Revista Brasileira de Ciências da Amazônia. 2014;**3**:5-28

[22] MMA - Ministério do meio ambiente. Manual para implantação de compostagem e de coleta seletiva no âmbito de consórcios públicos. Brasília- DF; 2010

[23] IBGE. Pesquisa Nacional de Saneamento Básico. Coordenação de População e Indicadores Sociais. Rio de Janeiro. 2010

[24] Saito ML. Documentos 64. O Uso do Lodo de Esgoto na Agricultura: precauções com os contaminantes orgânicos. [s.l: s.n.]

[25] CONAMA. Resolução n. 357 - 2006. Define critérios e procedimentos, para o uso agrícola de lodos de esgoto gerados em estações de tratamento de esgoto sanitário e seus produtos derivados, e dá outras providências. Brazil; 2006

[26] IPEA. Diagnóstico dos Resíduos Orgânicos do Setor Agrossilvopastoril e Agroindústrias Associadas. Brasília: Instituto de Pesquisa Econômica Aplicada – IPEA; 2012. p. 134

[27] Vazquez GH, Bortolin R, Vanzela LS, Bonini C dos SB, Bonini Neto A. Uso de fertilizante organofosfatado e torta de filtro em cana-planta. Brazilian Journal of Biosystems Engineering. 2015;**9**:53-64

[28] Charlton A, Sakrabani R, Tyrrel S, Casado MR, McGrath SP, Crooks B, Cooper P, Campbel CD. Long-term impact of sewage sludge application on soil microbial biomass: An evaluation using meta-analysis. Environmental Pollution. 2016;**219**:1021-1035

[29] Vieira RF, Pazianotto RAA. Microbial activities in soil cultivated with corn and amended with sewage sludge. Springer plus. 2016;**5**:1844-1859

[30] Kirchmann H, Börjesson G, Kätterer T, Cohen Y. From agricultural use of sewage sludge to nutrient extraction: A soil science outlook. Ambio. 2017;**46**:143-154

[31] Ren T, Wang J, Chen Q, Zhang F, Lu S. The effects of manure and nitrogen fertilizer applications on soil organic carbon and nitrogen in a high-input cropping system. PLS One. 2014;**9**:1-11

[32] Moretti SML, Bertoncini EI, Abreu-Junior CH. Composting sewage sludge with green waste from tree pruning. Scientia Agricola. 2015;**72**:432-439

[33] Aishah RM, Shamshuddin J, Fauziah CI, Arifin A, Panhwar QA. Phytoremediation of copper and zinc in sewage sludge amended soils using Jatropha curcas and Hibiscus cannabinus. Journal of the Chemical Society of Pakistan. 2016;**38**:1230-1243

[34] Diniz ICC, Matos AT, Borges AC, Aquino JMGL, Matos MP. Degradation of sewage sludge compost disposed on the soil. Engenharia Agrícola. 2016;**36**:822-829

[35] Boeira RC, Souza MD. Estoques de carbono orgânico e de nitrogênio, pH e densidade de um Latossolo após três aplicações de lodos de esgoto. Revista Brasileira de Ciência do Solo. 2007;**31**:581-590

[36] Liu H, Gao D, Chen T, Cai H, Zheng G. Improvement of salinity in sewage sludge compost prior to its utilization as nursery substrate. Journal of the Air & Waste Management Association. 2014;**64**:546-551

[37] Lim SR, Schoenung JM. Human health and ecological toxicity potentials due to heavy metal content in waste electronic devices with flat panel displays. Journal of Hazardous Materials. 2010;**177**:251-259

[38] Alloway BJ. Heavy Metals in Soils. New York: John Wiley & Sons; 1990. p. 339

[39] Alloway BJ. Heavy Metals in Soils. London: Blackie Academic; 1995. 368 p

[40] Kabata-Pendias A, Pendias H. Trace Elements in Soils and Plants. 3rd ed. Florida: Boca Raton, CRC Press; 2001. p. 413

[41] Bruins MR, Kapil S, Oehme FW. Microbial resistance to metals in the environment. Ecotoxicology and Environmental Safety. 2000;**45**:198-207

[42] Alloway BJ, Ayers DC. Chemical Principles of Environmental Pollution. 2nd ed. Boca Raton: CRC Press; 1996. p. 395

[43] Goyer RA. Toxic effects of metals. In: Klaassen CD, editor. Casarett & Doull's Toxicology: The Basic Science of Poisons. New York: McGraw Hill; 1996. pp. 691-736

[44] IPCS - International Programme on Chemical Safety. Environmental Health Criteria, 224 - Arsenic and Arsenic Compounds. Geneva: World Health Organization (WHO) [Internet]; 2001. Available from: http://www.inchem.org/documents/ehc/ehc/ehc224.htm [Accessed: February 15, 2017]

[45] IPCS - International Programme on Chemical Safety. Environmental Health Criteria, 134 - Cadmium. Geneva: World Health Organization (WHO) [Internet]. 1992. Available from http://www.inchem.org/documents/ehc/ehc/ehc134.htm [Accessed: February 03, 2017]

[46] IPCS - International Programme on Chemical Safety. Environmental Health Criteria, 61 - Chromium. Geneva: World Health Organization (WHO) [Internet]; 1988. Available from: http://www.inchem.org/documents/ehc/ehc/ehc61.htm [Accessed from: February 15, 2017]

[47] USEPA - United States Environmental Protection Agency. Reviews of the Environmental Effects of Pollutants. III. Chromium. Washington: USEPA; 1978. p. 285

[48] Quináglia GA. Estabelecimento de um protocolo analítico de preparação de amostras de solo para determinação de metais e sua aplicação em um estudo de caso [Dissertação]. São Paulo: Faculdade de Saúde Pública: Universidade de São Paulo; 2001

[49] Eleutério L. Diagnóstico da situação ambiental da cabeceira da Bacia do rio Doce, MG, no âmbito das contaminações por metais pesados em sedimentos de fundo [Dissertação]. Escola de Minas: Universidade Federal de Ouro Preto- Ouro Preto; 1977

[50] ATSDR - Agency for Toxic Substances and Disease Registry. Toxicological Profile for Lead. Atlanta: U. S. Department of Health and Human Services, Public Health Service; 1999

[51] ATSDR - Agency for Toxic Substances and Disease Registry. Case studies in Environmental Medicine – Lead Toxicity. Atlanta: U. S. Department of Health and Human Services, Public Health Service; 1992

[52] IPCS - International Programme on Chemical Safety. Environmental Health Criteria, 165 – Inorganic Lead. Geneva: World Health Organization (WHO) [Internet]; 1995. Available from: http://www.inchem.org/documents/ehc/ehc/ehc165.htm [Accessed: February 15, 2017]

[53] Vega-Dienstmaier JM, Salinas-Piélago JE, Gutiérrez-Campos MR, Mandamiento-Ayquipa RD, Yara-Hokama MC, Ponce-Canchihuamán J, Castro-Morales J. Lead levels and cognitive abilities in peruvian children. Revista Brasileira de Psiquiatria. 2006;**28**:33-39

[54] Bellinger DC. Interpreting the literature on lead and child development: The neglected role of the experimental system. Neurotoxicology and Teratology. 1995;**17**:201-212

[55] CETESB - Companhia de Tecnologia de Saneamento Ambiental. Método de Ensaio – Salmonella: Isolamento e Identificação. 1993. Norma L5 218

[56] Rodrigues RB, da Silva Júnior TAF, Maringoni AC. Efeito da aplicação de lodo de esgoto na severidade da murcha-de-curtobacterium em feijoeiro. Summa Phytopathologica. 2006;**32**:82-84

[57] CETESB - Companhia de Tecnologia de Saneamento Ambiental. Séries Manuais. Avaliação de Desempenho de Estações de Tratamento de Esgotos. São Paulo: 1991. p. 37

[58] Leal RMP. Ocorrência e comportamento ambiental de resíduos de antibióticos de uso veterinário [Tese]. Piracicaba: Centro de Energia Nuclear na Agricultura, Universidade de São Paulo; 2012. DOI: 10.11606/T.64.2012.tde-04102012-145438. Acesso em: 08 de fevereiro de 2017

[59] Abreu Junior CH, Boaretto AE, Muraoka T, Klehl JC. Uso agrícola de resíduos orgânicos potencialmente poluentes: Propriedades químicas do solo e produção vegetal. In: Tópicos em Ciência do Solo. 4th ed. Viçosa: Sociedade Brasileira de Ciência do Solo; 2005. pp. 491-470

[60] Alfaro MR, Montero A, Ugarte OM, Nascimento CWA, Accioly AMA, Biondi CM, Silva YJAB. Background concentrations and reference values for heavy metals in soils of Cuba. Environmental Monitoring and Assessment. 2015;**187**:4198

[61] de Almeida Júnior AB, do Nascimento CWA, Biondi CM, de Souza AP, do Barros FMR. Background and reference values of metals in soils from Paraíba State. Brazil. Revista Brasileira de Ciência do Solo. 2016;**40**:1-13

[62] Nascimento CWA, Biondi CM. Valores orientadores da qualidade do solo para metais. In: Tópicos em Ciência do Solo. 9. ed. Viçosa: Sociedade Brasileira de Ciência do Solo; 2015. p. 290

[63] Conama - Conselho Nacional Do Meio Ambiente. Resolução 420, de 28 de dezembro de 2009 [Internet]. 2009. Available from: http://www.mma.gov.br/port/conama/legiabre.cfm?codlegi=620 [Accessed: February 08, 2017]

[64] Chen M, Ma LQ, Harris WG. Baseline concentrations of 15 trace elements in Florida Surface Soils. Journal of Environmental Quality. 1999;**28**:1173-1181

[65] Biondi CM, Nascimento CWA, de Neta A BF, Ribeiro MR. Teores de Fe, Mn, Zn, Cu, Ni e Co em solos de referência de Pernambuco. Revista Brasileira de Ciência do Solo. 2011;**35**:1057-1066

[66] Cetesb - Companhia Ambiental do Estado de São Paulo. Decisão de diretoria n°256/2016. São Paulo: CETESB; 2016:55-56

[67] Amaral AS, Defelipo BV, Costa LM, Fontes MPF. Liberação de Zn, Fe, Mn e Cd de quatro corretivos da acidez e absorção por alface em dois solos. Pesquisa Agropecuária Brasileira. Brasília.1994;**29**:1351-1358

[68] Boechat CL, da Silva PSP. Chemical characterization to evaluate the agricultural potential use of organic wastes generated by industrial and urban activity. African Journal of Agricultural Research. 2012;**7**:3939-3944

[69] Boechat CL, Ribeiro MO, Ribeiro LO, Santo SJAG, Accioly AMA. Lodos de esgoto municipal e industrial no crescimento inicial e qualidade de mudas de Pinhão-manso. Bioscience Journal. 2014;**30**:782-791

[70] Boechat CL, Ribeiro MO, Bomfim MR, Bittencourt NS, Accioly AMA, Santos JAG. Sewage sludges in physic nut seedlings macronutrient contents and chemical attributes of soil. Bioscience Journal. 2015;**31**:1378-1387

[71] Prado RM, Coutinho ELM, Roque CG, Villar MLP. Avaliação da escória de siderurgia e de calcários como corretivos da acidez do solo no cultivo da Alface. Pesquisa Agropecuária Brasileira. 2002;**37**:539-546

[72] Prado RM, Corrêa MCM, Cintra ACO, Natale W. Resposta de mudas de goiabeira à aplicação de escória de siderurgia como corretivo de acidez do solo. Revista Brasileira de Fruticultura. 2003;**25**:160-163

[73] Malavolta E, Pimentel-Gomes F, Alcarde JC. Adubos & adubações. São Paulo: NBL Editora; 2002. p. 200

[74] Souza JL, Resende P. Manual de Horticultura Orgânica. 2nd ed. Viçosa: Aprenda Fácil; 2006. p. 843

[75] Santana CT, Adalberto SD, Santo ML, Menezes CB. Desempenho de cultivares de alface americana em resposta a diferentes doses de torta de filtro. Revista Ciência Agronômica. 2012;**43**:22-29

[76] Boechat CL, Ribeiro MO, Santos JAG, Accioly AMA. Mineralizable nitrogen of organic wastes and soil chemical changes under laboratory conditions. Communications in Soil Science and Plant Analysis. 2014;**45**:1981-1994

[77] Araújo EF, de Arauco AMS, de Lacerda Junior JJ, Ratke RF, Medeiros JC. Growth and nutrient balance of Enterolobium contortsiliquum seedlings with addition of organic substrates and wastewater. Brazilian Journal of Forestry Research. 2016;**36**:169-177. DOI: 10.4336/2016.pfb.36.86.1135

Chapter 6

Home Composting Using Facultative Reactor

Sandro Xavier de Campos, Rosimara Zittel,
Karine Marcondes da Cunha and
Luciléia Granhern Tavares Colares

Additional information is available at the end of the chapter

http://dx.doi.org/10.5772/intechopen.69429

Abstract

Concerns with the final destination of organic solid waste (OSW) generated in rural areas originate from the possibility of this waste harming the environment, in addition to producing bad smell and attracting pests, when improperly disposed of in the soil. In this sense, composting might be a suitable way of dealing with this residue. This chapter presents the advantage of treating rural OSW through composting in reactors. Facultative reactors present the advantage of not requiring handling or large areas for the waste processing, and they do not generate bad smell and do not attract pests, which represent common drawbacks of the conventional windrow composting process. The final product of this composting process can be used as fertilizer for crops, resulting in the economy, since commercial fertilizers do not have to be bought. Works carried out by the Analytical and Environmental Chemistry Research Group at the State University of Ponta Grossa—Brazil have reported important results regarding the use of facultative reactors with different OSW mixtures. From the monitoring of physical, chemical, biological and spectroscopic parameters, it was seen that composting in facultative reactors produced stable compost matured in a short period of time.

Keywords: facultative reactor, OSW, monitoring, characterization, spectroscopic, physico-chemical

1. Introduction

The final destination of solid residues originated in agriculture as well as other activities has been a pressing concern in the contemporary society, due to its negative impact on the environment. In the whole world, 1.3 billion tons of solid waste are generated every year according to the Organization for Economic Cooperation and Development (OECD). It is estimated that 50% of this waste is produced by the 34 richest countries. Brazil, despite being considered an underdeveloped country, occupies the third position among the countries that generate the most waste in the world (220 million tons/year, 1.2 kg/inhabitant/day) [1].

The Brazilian agroindustry is the largest productive sector in the country. According to data by the United Nations (UN), Brazil will be the greatest food exporter between 2015 and 2024. Such a great food production places Brazil among the largest waste producers also in the rural area [1].

According to the Applied Research Institute (IPEA/Brazil) [2], about 291 million tons of waste are generated by the agroindustry every year. From this, 51% is organic material coming from solid waste from sugar cane, rice, soybean, corn, beans, wheat, coffee and cocoa crops, along with the growth of fruits, such as orange, banana, coconut and grapes. This organic solid waste (OSW) is usually buried, burnt or simply disposed on the soil far from households (but many times close to rivers), generating negative environmental impact [3, 4]. However, if properly separated and treated, this waste can contribute to the reduction of environmental problems originated from its improper disposal. The use of composting methods that can transform this OSW into stable and mature organic matter in the shortest period possible is the most suitable way of managing such residue.

This chapter presents the possibility of treating OSW through composting in facultative reactors.

2. OSW treatment in reactors

In the rural area, OSW from different sources, both vegetable and animal origin, is generated inside and outside the farms. Proper management of this residue might result in benefits related to the prevention of river and soil pollution; reduction in chemical fertilizer use and crop diseases. Therefore, the collection and composting of this OSW might constitute proper technology, to be used by individuals or in association with other rural producers, aiming at the agriculture technical-scientific advancement [5]. Up to now, some emphasis has been given to home composting, which requires constant care such as moist content and material aeration control. In addition, studies have demonstrated that due to the requirement of constant handling, the composting process might present chemical and biological risks [5–7].

Taking these facts into consideration, the use of a reactor for the treatment process of OSW generated in rural areas might be suitable, since it represents a low-cost process which demands reduced workforce and small areas.

The composting in reactors is considered a promising technology when compared to conventional technologies of open systems such as windrows or piles, since it does not require revolving the composting mass and provides sufficient aeration to the mixture (with or without mechanical injection of air) to produce mature final compost. It does not produce the bad smell, leachate or pollutants. Also, it provides the control of physical and chemical parameters such as temperature and moist and can be used in different climatic seasons [8–10].

Recent studies employing reactors for OSW treatment have proved the efficiency of this technology when compared to the anaerobic digestion or incineration, for resulting in compostable to improve the physical and chemical conditions of the soil, preventing the emission of greenhouse effect gases (CO_2, NO, CH_4) [11]. In recent years, models of vertical and or horizontal reactors have been developed and adapted to OSW treatment [12, 13], and also several works have demonstrated composting processes in pilot-scale reactor systems, with rotating the drum and forced aeration [14–16].

Scale reactors between 10 and 300 liters involve different configurations that guarantee a process with favorable results, allowing the study of parameters such as temperature, moist, biological activity and oxygen content [17–19].

Taking that into consideration, some studies have emphasized that in order to retain heat and keep ideal temperatures, the reactor walls should be covered with thermal insulating material [14, 15, 20–22]. Several studies used OSW coming from vegetable, fruit, olive bagasse, grape bagasse and olive bagasse, waste which is widely found in rural areas. Due to the high moist of these substrates, in some experimental processes, output holes were adapted to drain the material and collect slurry [14, 22–24].

A study carried out by Fernández et al. [14], verified the influence of the granulometry in temperature and moist variation in a reactor system for the treatment of sewage sludge coming from an effluent treatment station (ETS), combined with carbonaceous materials with different particle sizes. The authors concluded that the best results were obtained from the ETS sludge and wood chips (5–15 mm). The use of wood chips as volume agents provides the mixture with better aeration and moist levels below 65%, in addition to favoring the achievement of temperatures close to 70°C, ensuring the elimination of pathogens.

Paradelo et al. [22] investigated the efficacy of different volume agents and studied the mixtures of food waste (raw and cooked vegetables and fruit) with the addition of different OSW produced in rural areas such as hay, wheat straw and wood chips. The composts were placed in a vertical cylindrical barrel of 30-liter volume for 2 months. The barrel was placed on wood blocks to favor aeration and elimination of the leachate produced. The authors concluded that the wood chip experiment produced a less decomposed material, due to the larger particle size (25 mm); however, it also presented parameters favorable to stabilization.

In another study by Li et al. [21], a reactor was built with plastic containers of different sizes, a 10 L container, punctured and with external output, encased in a 15 L container. The space between the reactors was 5 mm each side (filled with foam) and 20 mm in the lower part. At the bottom of the bucket, gravel was placed and that surface was covered with a plastic screen. The composting used grape bagasse and sawdust. The treatment was carried out with the lid of the system closed. The authors concluded that the cellulose degradation occurred within the three initial months; however, the variation of the C/N ratio from 30/1 (initial) to 28/1 (final) was low, and the temperature did not reach the thermophilic phase.

Composting in rotating drums with an aeration system and 250 L capacity has been widely studied [23]. Different mixtures of OSW produced in rural areas were used such as grass, vegetable residue, bovine manure and sawdust. The authors concluded that the mixture of bovine manure, vegetable residue and sawdust resulted in a stabilized compound and agitation guaranteed aeration and uniformity to the mixture, preventing the generation of bad smell.

Iyengar and Bhave [25] investigated composting in different kinds of reactors (aerobic, anaerobic and facultative) in laboratory scale, using OSW from rural areas such as bovine manure and straw. The facultative reactor presented anaerobic lower regions and aerobic upper regions, due to the distribution of the layers inside the reactor. The authors concluded that the aerobic and facultative reactor systems produced a more stable compost.

For the OSW composting process in reactors to be considered efficient, it is necessary to monitor the different physical and chemical parameters, and the compost obtained can be defined as the stabilized and matured product.

The compost stability and maturity show the organic matter decomposition degree, and after mature, it can be used as a soil fertilizer, releasing nutrients necessary to plant growth [9]. Stability and maturity can be monitored by observing the physicochemical properties (temperature, moisture, pH, ash content and ratio C/N) along with spectroscopic (UV/Vis and IR) and biological (germination index) parameters [26–29].

In recent years, the Analytical Environmental and Sanitary Chemistry Research Group (QAAS, Brazilian abbreviation) of the State University of Ponta Grossa—Brazil has studied the treatment of different OSW in a facultative reactor system [9, 30, 31]. Thus, this work presents an extensive study of the OSW treatment in the facultative reactor, with great results obtained through conventional (moisture, pH, temperature and C/N ratio), spectroscopic techniques (UV/Vis and Infrared-IR) and germination index.

2.1. Facultative reactor

The facultative reactor (**Figure 1**) was constructed from a copper and zinc cylindrical metal container with a capacity of 200 L (diameter 600 mm and height 700 mm). On the reactor was designed a plastic cover of 300 mm of diameter × 60 mm of height that contains 120 holes of 5 mm to allow a gas exchange. At the bottom of the reactor was coupled a liquid collection system [9].

2.2. Composting using facultative reactor

The experiments were carried out at the State University of Ponta Grossa, where five reactors were assembled, and the composting process was studied for 180 days. The reactors were installed under shelter (shed) protected from the rain. For the experiment, the following OSW was used: home organic waste (HOW), wood waste such as sawdust (WWS) and chips (WWC) and smuggled cigarette tobacco (SCT) seized by the Brazilian Federal Revenue and donated to the study. The combination of OSW used to assemble the five reactors is in **Table 1**. The mixture was distributed in 200 mm layers in the reactors. **Table 2** presents the physical and chemical characteristics of the OSW used in the experiment.

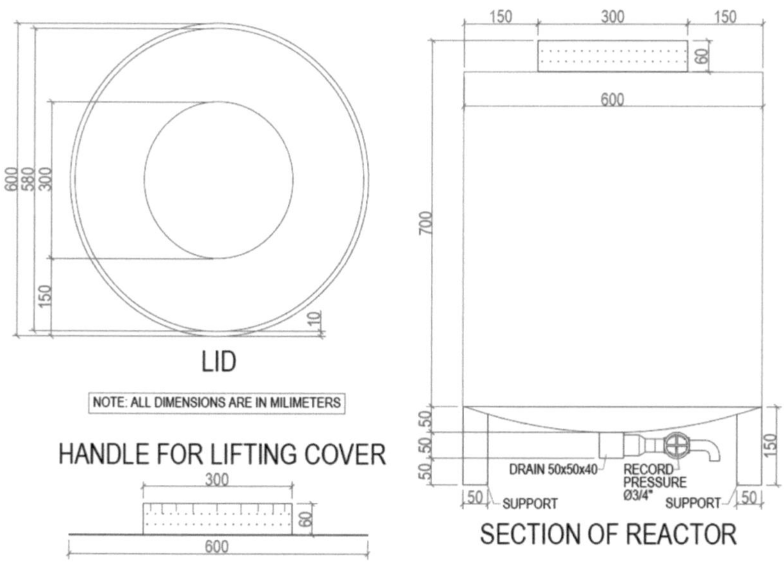

Figure 1. Schematic design of the facultative reactor.

	DOW		SCT		WC		S		Total
	(%)	(kg)	(%)	(kg)	(%)	(kg)	(%)	(kg)	(kg)
R 1	70	157.2	10	2.62	20	8.7	–	–	168.2
R 2	60	116.4	20	4.84	20	8.0	–	–	129.3
R 3	40	86.3	40	10.1	20	8.3	–	–	104.8
R 4	70	105.6	10	1.69	–	–	20	2.9	110.2
R 5	40	57.6	40	6.7	–	20	2.8	67.1	

Source: Zittel [31].

Table 1. Combination of the substrates, in proportion and mass, for the accomplishment of the experiment.

Parameters	DOW	SCT	WRS	RL
pH	5.5	7.2	6.4	6.0
C (%)	36.15	36.85	44.1	44.1
N (%)	3.04	3.20	0.28	0.28
Initial C/N ratio	11.89	11.15	157.5	157.5
Granulometry (mm)	20–50	> 1	> 1	20–40
Moisture content (%)	65	15,3	12	13

Source: Zittel [31].

DOW—domestic organic wastes; SCT—smuggled cigarette tobacco; WRS—wood residues sawdust; RL—Wood Residues Chip.

Table 2. Initial physical and chemical characteristics of the solid residues used in the experiment.

2.2.1. Home organic waste (HOW)

The average of six samplings was carried out to characterize the HOW, and a quartering process was performed three times a week according to the Technical Norms Brazilian Association—ABNT-NBR 10007/2004 [32]. Initially, the homogenized sample was divided into four parts, and two opposite quarters were selected, which were homogenized again. The quartering procedure was carried out again with the samples collected, selecting one of the remaining quarters to represent the waste characterization (**Figure 2**).

2.2.2. Wood waste-sawdust (WWS) and chips (WWC)

The WWS used in the experiment had a diameter smaller than 1 mm, and the WWC diameter was between 20 mm and 40 mm, as shown in **Figure 3**.

2.2.3. Smuggled cigarette tobacco (SCT)

The SCT used in the experiment was provided by the Brazilian Federal Revenue (Ponta Grossa—Pr unit) after having sized the material. Different batches of this material were used to assemble the reactors. The cigarette filter was separated from the tobacco portion and ground in a commercial grinder shown in **Figure 4**.

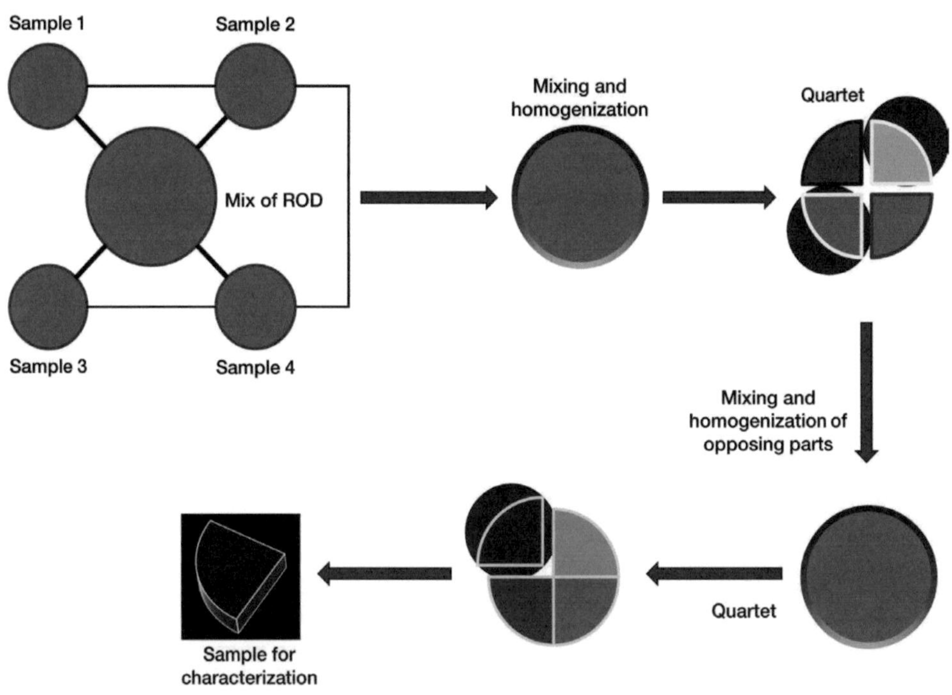

Figure 2. Quarantine process of the RODs used in the experiment.

Figure 3. Wood waste used in the experiment. (A) Sawdust wood residue (RMS) and (B) residual wood chippings (RML).

Figure 4. Equipment used for grinding tobacco and paper wrapping.

After ground, the tobacco was separated from the wrapping paper around it using a 6 mm × 6 mm mesh sieve, resulting in a powder with particle smaller than 1 mm. The process to separate the tobacco from the cigarette filter and wrapping paper can be seen in **Figure 5**.

2.3. Sampling

To monitor the composting process, samples were collected every fortnight, from random points inside the reactors. The sample collection was carried out on the 1st day and in periods of 15, 30, 45, 60, 75, 90, 105, 120, 135, 150, 165, 180 days. In each collection, about 100 g of sample was removed from each reactor. These samples were homogenized, and the portions needed for the analyses were separated. To monitor de composting process and characterization of the final compost, the analyses carried out were: elemental (C/N), temperature, moist, pH, seed germination index (SGI) and spectroscopic (UV/VIS and Infrared).

2.4. Physicochemical analyses

2.4.1. Temperature

The temperature was measured three times a week using a portable digital thermometer. The monitoring was carried out on the first day of the organic material deposition inside the reactor and throughout the process, at three different points: surface, middle and bottom.

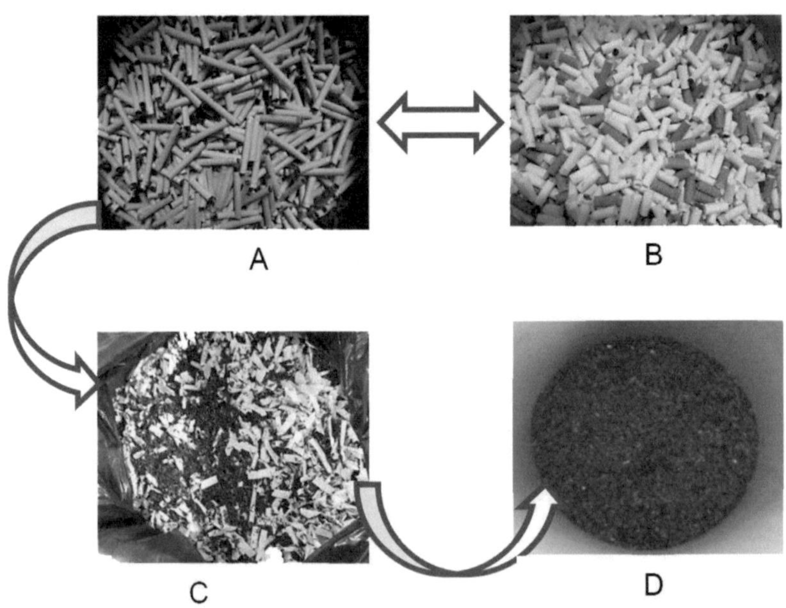

Figure 5. Separation steps of the filter and the paper wrapper of the cigarette used in the experiment. (A) Cigarette with filter; (B) filter separation; (C) shredded tobacco and paper wrapping; and (D) cigarette tobacco residue after sieving.

2.4.2. *Moist*

To control moist content, readings of the mixture in the reactors were performed every fortnight, at the surface and bottom of the reactor, using the combined digital thermometer and hygrometer device.

2.4.3. *pH*

For the pH determination, the sample was diluted in a calcium chloride solution, following the official method by the Brazilian Agriculture, Livestock and Supply Ministry [33]. The use of a calcium chloride solution allows the hydrogen ions to get free from the colloid surface, resulting in pH values which are moderately lower when compared to those found when the sample is diluted in water [34].

2.4.4. *Elemental chemical analysis*

The elemental chemical analysis was carried out using the Elemental Analyzer. The C/N ratio was obtained from the ratio between the percentage values of C and N in the sample.

2.4.5. *Germination index (GI)*

GI was performed by adapting the technique proposed by the Hong Kong Organic Resource Centre [35]. About 20 g of the fresh compost was added to 200 mL of distilled water and stirred for 2 h. Next, the solution was placed in the centrifuge at 9000 rpm for 25 min at 4°C. The GI was carried out in triplicate in the periods of 90, 120, 150 and 180 days. GI was performed by placing 5.0 mL compost extract on each petri dish. Next, 10 watercress seeds (*Lepidium sativum*) were evenly distributed on each dish and incubated for 72 h in dark room and temperature between 22 and 27°C. After 72 h, the number of germinated seeds was recorded, as well as the main root length, so that the GI calculation could be done [34] (Eqs. (1)–(3)). The percentage of germinated seeds of each sample as calculated from the triplicate weighted average. A control test was performed using 5.0 mL distilled water substituting the compost extract.

Relative seed germination percentage

$$RSG(\%) = \frac{\text{Weighted average of the no of germinated seeds in each extract} \times 100}{\text{Weighted average of the no of germinated } - \text{ seeds in the control}} \tag{1}$$

Relative root growth

$$RRG(\%) = \frac{\text{Weighted average of root length in each extract} \times 100}{\text{Weighted average of root lenght in the blank}} \tag{2}$$

Germination index (GI)

$$GI = \frac{RSG \times RRG}{100} \tag{3}$$

2.5. Spectroscopic analysis

2.5.1. *Visible ultraviolet region (UV–Vis) molecular spectroscopy*

For the absorption analyses in the UV–Vis, 10 mg of the sample was dissolved in 10 mL sodium bicarbonate solution ($NaHCO_3$) 0.05 mol. L-1, at the E_2/E_3 (wavelength absorbance 280 nm and 365 nm) and E_4/E_6 (ratio between the absorbances 465 and 665 nm).

2.5.2. *Infrared region Absorption Spectroscopy (IR)*

For the IR analyses, pellets were prepared with 1.0 mg sample and 100 mg KBr. The spectra were obtained in the band from 400 cm^{-1} to 4000 cm^{-1} [36].

3. Results

3.1. Physicochemical parameters

Figure 6 presents the results of physical and chemical analyses of the OSW mixtures used in the composting with the facultative reactor.

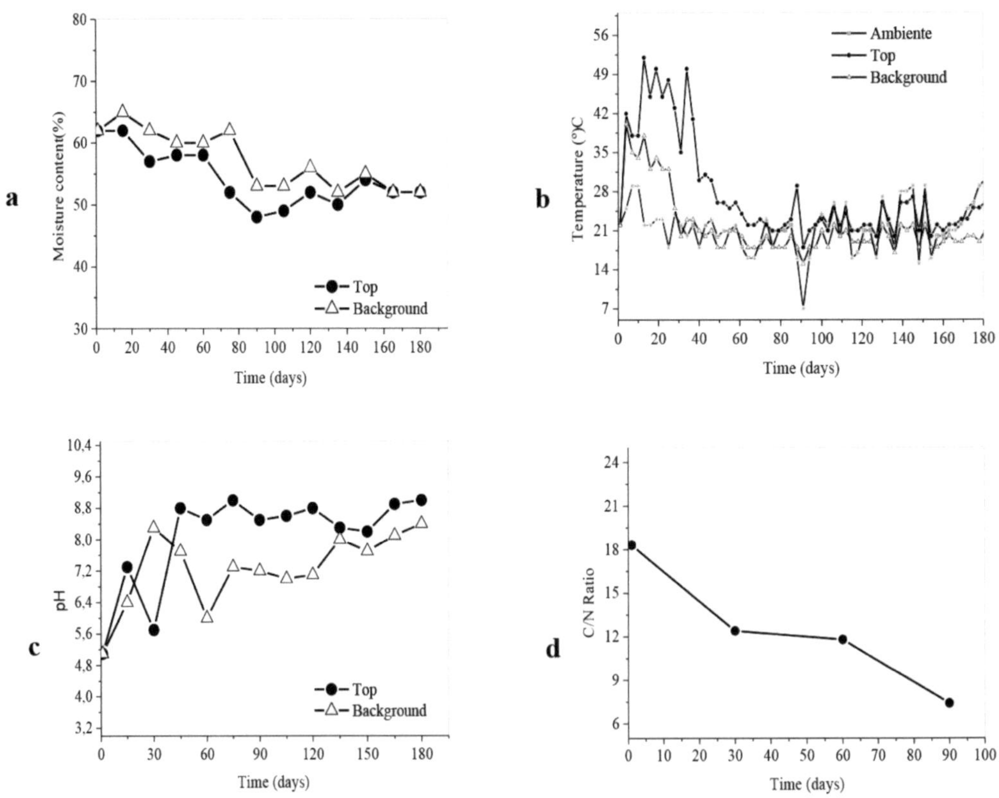

Figure 6. Control of moisture (a), temperature (b), pH (c) and C/N ratio (d) during the composting process of food residues (40%), cigarette tobacco (40%) and sawdust (20%) in the optional reactor system.

The moist (**Figure 6a**) in both layers (top and bottom) was seen to be kept between 40 and 70% throughout the process, and from the 130th day on, there was no variation of this value at the top or at the bottom of the mass, indicating the compost stability. Regarding pH (**Figure 6b**), a reduction was seen in the first 15 days in the two layers of the mixture, while, after the 80-day period, the values were in the alkaline band. From the 130th day on, the values were seen to vary from 7.5 to 9, indicating the final compost stability. The variation in pH values is due to the existence of different groups of microorganisms which are characteristic of the aerobic (bacteria, yeast, and fungi) and anaerobic (bacteria) activities. **Figure 6c** shows that the thermophilic temperature remained for a period of 40 days, which was efficient to eliminate pathogens. From the 100th day on, temperatures close to the ambient were observed, confirming the compost stability.

The C/N ratio is a parameter that evaluates the compost maturity. According to the results presented in **Figure 6d**, the ratio was seen to decrease throughout the process, indicating that the initial materials, rich in nitrogen, were transformed into inorganic compounds such as nitrates (NO_3) and nitrites (NO_2).

3.2. Germination index (GI)

Recently, the phytotoxicity evaluation through the germination index (GI) has been used as one of the parameters to evaluate the compost maturity. The application of unstable and immature compost to the soil might result in competition for oxygen by the plant roots and the microbial mass. The continuous decomposition of immature materials might cause anaerobic conditions in the soil, resulting in the production of nitrite (NO_2) and sulfuric acid (H_2S) [37]. Immature composts might contain high levels of organic acids, high C/N ration, extreme pH values, high salinity content and high ammonium (NH_4) concentrations, which inhibit seed germination and root and plant growth [38].

When the GI increases along the process, one can assume that there is a reduction in phytotoxic substances and that the organic composts are reaching maturity, being enriched with nutrients and humic substances [39].

Figure 7 presents the GI evolution in OSW composting with the facultative reactor.

Results revealed that from the 90th day on the compost was free of toxicity and reached SGI values above 80%.

3.3. Visible ultraviolet region (UV–Vis) spectroscopic analyses

The formation of humic substances is a parameter used to evaluate the maturity of the final compost obtained in OSW composting with the facultative reactor. Therefore, the humification parameter among some ratios of absorbance is widely used. The main organic material absorption bands occur in the region from 200 to 400 nm. In composting studies, the compost UV/Vis analyses result in ratios between some absorbance. The E_2/E_3 ratios (ratio between the absorbance 280 and 365 nm) provide the relation between humified and non-humified groups [40], while the E_4/E_6 ratios (ratio between the absorbance 465 and 665 nm) are used to indicate the condensation degree and aromatic constituents during composting and may be seen as humification index or compost maturity [40]. These ratios usually reduce with the increase in simple and double chemical bond conjugations characterizing the formation of humic substances through the condensation of aromatic rings of greater molecular weight [41, 42]. The

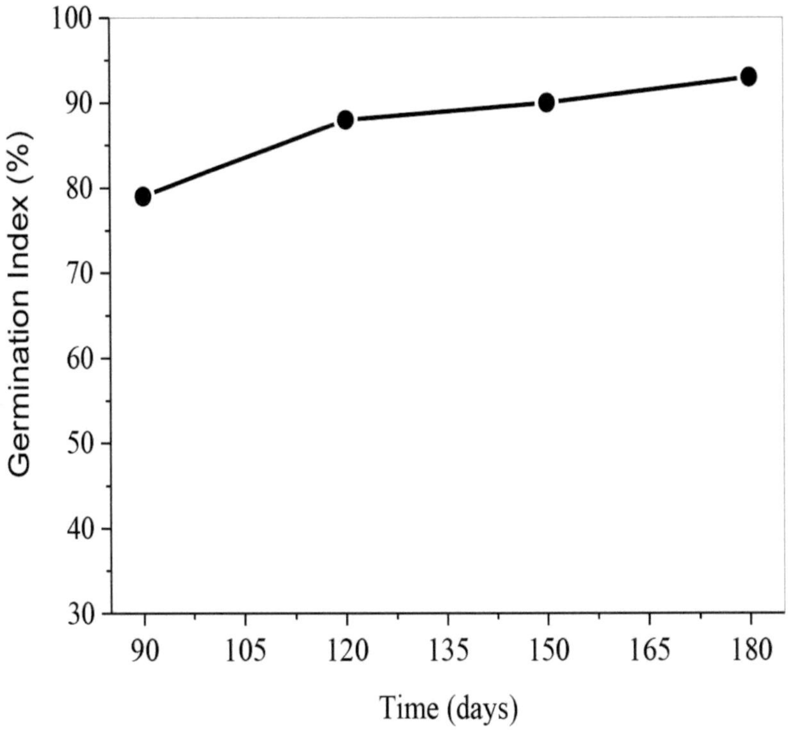

Figure 7. Seed germination index during composting process of food residues (40%), cigarette tobacco (40%) and brush (20%).

relations E_2/E_3 and E_4/E_6 with values lower than 5.0 indicate that there was the formation of mature compost due to the increase in the aromatic groups [43].

Figure 8 presents the E_2/E_3 and E_4/E_6 ratio curve during the OSW composting process with the facultative reactor.

The E_2/E_3 and E_4/E_6 ratios were seen to decrease during the process reaching values below 5. These results indicated that the composting in reactors produced matured final compost probably due to the high degree of humic acids present in the humic substance structure.

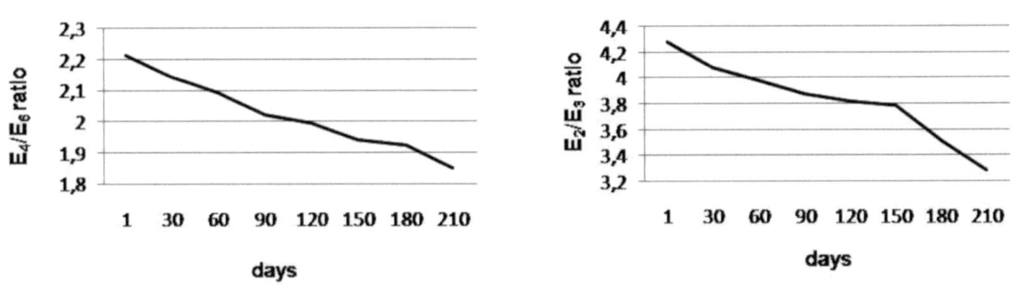

Figure 8. Curve of E_2/E_3 ratio (absorbance between 250 and 365 nm) and E_4/E_6 absorbance between 465 and 665 nm during the process of composting food residues and sawdust.

3.4. Fourier transform infrared spectroscopy (FT-IR)

The infrared spectroscopy (IR) has been used to evaluate the transformation of organic matter at the different stages of the composting process [44]. The increase and reduction in peaks reveal the decomposition of immature residue and the formation of new compounds [45].

The main absorption peaks of the IR spectrum and their respective attributions are seen in **Table 3** [44, 46].

The IR spectrum of the OSW composting process with the facultative reactor in different periods is presented in **Figure 9**.

After analyzing the main absorption peaks in the different periods of the composting process, it was seen that around 3400 cm^{-1} a band characteristic of OH alcohols, phenols and carboxyl acids occurred and another with aliphatic groups was seen around 2928 cm^{-1}. These bands decreased along the process, due to the degradation of these groups characteristic of non-humic substances (carbohydrates and fat acids) and the formation of humic substances due to the microbiological action during the process [39, 41].

The peaks between the absorbance 1700–1750 cm^{-1} conferred the C=O stretching vibration, of carboxyl acid groups, ketones, and aldehydes, characterizing structures made of the non-ionized

Wave number (cm^{-1})	Assignments
3550–3300	H–OH vibration elongation (phenol groups, alcohols and carboxylic acids); N–H (amides and amines)
2920	C–H vibration elongation of aliphatic structures
1715–1750	C=O vibration elongation of COOH of carboxylic acids and ketones
1620–1660	C=O vibration elongation of primary amides
	C=O vibration elongation of ketones, acids and quinones
	Vibration elongation C=C of aromatics
1505–1560	Vibration elongation C=C of lignin aromatics; N–H and C=N of amine and secondary amide
1460	Vibration elongation C=C of aromatics, O–H of phenols
1375–1390	Vibration elongation COO, C–O of carboxylic acids and/or carbonates and nitrates and deformation of vibrations O–H of phenols
1220–1250	C–H and OH vibration deformation of carboxyl groups, C–O–C aromatic ether and N–H of secondary amides
1120 and 1030–1050	C–C vibration elongation of aliphatic
	C–O vibration elongation of polysaccharides, C–O of aromatic-cycle ether

Source: Zittel [31].

Table 3. Peaks of absorption of the spectrum of IR and its attributions.

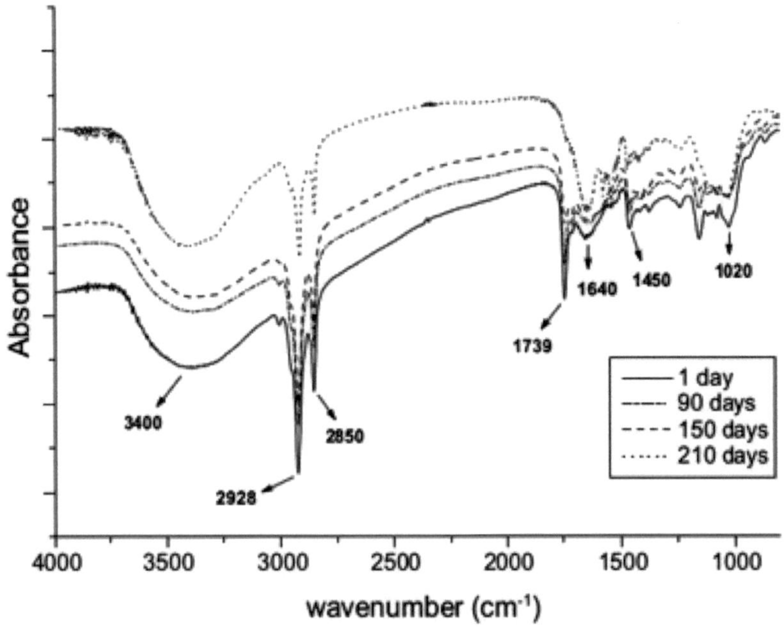

Figure 9. IV spectrum of food waste samples and sawdust in different periods of the composting process.

carbonyl. The spectrum region between 1600 and 1660 cm^{-1} is related to the intensity of the C=O groups of ionized carboxyl (COO^{-}) and conjugated to the aromatic ring and humic acids [39, 47].

The formation of a sharp peak between 1375 and 1390 cm^{-1} in 180 days of the process might be ascribed to the deformations of O=H of phenolic groups, present in the humic substance structure, while the vibrations in the 1010–1035 region indicated polysaccharide, C=O, stretching [41, 48].

The increase in the intensity of peak absorption at 1450 and 1390 cm^{-1} indicated the presence of oxidation reactions, with the formation of carboxyl and carbonate acids [46, 47, 49].

The IR spectrum presented absorption peaks which indicated the transformation of organic matter and matured compost production, characterizing the increase in the humic acids during the process.

4. Conclusions

From the results obtained, it was possible to conclude that the facultative reactor presented the advantages of being a low-cost system, not attracting vectors, enabling moisture and temperature control without the need for handling, besides allowing the treatment of several organic waste characteristic of rural areas. Phytotoxicity tests showed that the compost reached maturity over a period of 90 days. The spectroscopic analyzes showed that degradation of compounds of simpler structure and the humification of the organic matter occurred. Thus, the compost obtained, with a substantial richness of stabilized organic matter and absence of toxicity, may be considered as an organic fertilizer. Finally, this study led to the conclusion that the facultative reactor proposed can be a promising technology to manage organic solid residue in rural areas.

Acknowledgements

We would like to thank CAPES, which was especially helpful at all stages of the research providing the researchers with scholarships.

Author details

Sandro Xavier de Campos[1]*, Rosimara Zittel[1], Karine Marcondes da Cunha[2] and Luciléia Granhern Tavares Colares[3]

*Address all correspondence to: campos@uepg.br

1 State University of Ponta Grossa, Brazil

2 Federal Institute of Paraná, Jaguariaíva Campus, Brazil

3 Federal University of Rio de Janeiro, Brazil

References

[1] OECD/FAO. OECD-FAO Agricultural Outlook 2016-2025. OECD Publishing, Paris. [Online]. 2016-2025 [cited 2017 março 25. Available from: http://dx.doi.org/10.1787/agr outlook-2016-en

[2] Instituto de Pesquisas Aplicadas—IPEA. Plano Nacional de Resíduos Sólidos: diagnóstico dos resíduos urbanos, agrosilvopastoris e a questão dos catadores. 2012

[3] Massukado LM. Desenvolvimento do processo de compostagem em unidade descentralizada e proposta de software livre para o gerenciamento municipal de resíduos sólidos domiciliares. Programa de pós graduação em ciências ambientais. 2008

[4] Zurbrügg C, Drescher S, Patel A, Sharatchandra HC. Decentralised composting of urban waste: An overview of community and private initiatives in Indian cities. Waste Management. 2004;**24**: 655-662

[5] Teves IC, Miller PRM. Compostagem: Ciência e prática para gestão de resíduos orgânicos. Rio de Janeiro: Embrapa Solos; 2009

[6] Gajalakshmi S, Abbasi AA. Solid waste management by composting: State of the art. Environmental Science Technology. 2008;**38**: 311-400

[7] Oliveira AMG, Aquini AM, Castro Neto MT. Bahia: Embrapa; 2005 [Online] [cited 2010 junho 20. Available from: https://www.embrapa.br/mandioca-e-fruticultura/publicacoes/circulares/circular_76.pdf

[8] Castilhos Junior AB, Pessin N, Fernandes F. Gerenciamento de Resíduos Sólidos Urbanos com Ênfase na Proteção de Corpos D'água: Prevenção, Geração e Tratamento de Lixiviados de Aterros Sanitários: Prosab–Rede Cooperativa de Pesquisas; 2006

[9] Campos SX, Ressetti RR, Zittel R. Monitoring and characterization of compost obtained from household waste an pine sawdust in a facultative reactor by conventional and spectroscopic analysis. Waste Management & Research. 2014;**32**: 1186-1191

[10] Tàtano F, Pagliaro G, Di Giovanni P, Floriani E, Mangani F. Biowaste home composting: Experimental process monitoring and quality control. Waste Management. 2016;**50**: 31&–38

[11] United States Environmental Protection Agency—US EPA. Types of Composting and UnderstandingtheProcessRetrievedfromandundertandingprocessinvessel.[Online];2016 [cited 28 November 2016. Available from: www.epa.gov/sustainable-management-food/types-composting

[12] Dong JD, Zhang YY, Wang YS, Wu ML, Zhang S, Cai CH. Chemometry use in the evaluation of the sanya bay water quality. Brazilian Journal of oceanography. 2010;**58**(4): 339-352

[13] Fernández J, Perez M, Romero LI. Effects of substrate concentration on dry mesophilic anaerobic digestion of organic fraction of municipal solid waste (OFSW). Bioresource Technology. 2008;**99**: 6075-6080

[14] Fernández FJ, Sánches-Arias V, Rodriguez L, Villaseñor J. Feasibility of composting combinations of sewage sludge, olive mil waste and winery waste in a rotary drum reactor. Waste Management. 2010;**30**: 1948-1956

[15] Rodrígues L, Cerrillo MI, García-Albiach V, Villaseñor J. Domestic sewage sludge composting in a rotary drum reactor: Optimizing the thermophilic stage. Bioresource Technology. 2012;**112**:284-291

[16] Villaseñor J, Rodríguez Mayor L, Rodríguez Romero L, Fernández FJ. Simulation of carbon degradation in a rotary drum pilot scale composting process. Journal of Environmental Management. 2012;**108**: 1-7

[17] Lashermes G, Barriuso E, Villio-Poitrenaud ML, Houot S. Composting in small laboratory pilots: Performance and reproducibility. Waste Management. 2012;**32**:271-277

[18] Mason IG, Milke MW. Physical modelling of the composting environment: A review. Part 1: Reactor systems. Waste Management. 2005;**25**: 481-500

[19] Yu H, Huang GH. Effects of sodium acetate as a pH control amendment on the composting of food waste. Bioresource Technology. 2009;**100**:2005-2011

[20] Kim KY, Kim HW, Han SK, Hwasng EJ, Lee CY, Shin HS. Effect of granular porous media on the composting of swine manure. Waste Management. 2008;**28**:2336-2343

[21] Li Q, Wang XC, Zhang HH, Shi HL, T.Hu, Ngo HH. Characteristics of nitrogen transformation and microbial community in an aerobic composting reactor under two typical temperatures. Bioresource Technology. 2013;**137**:270-277

[22] Paradelo R, Moldes AB, Barral MT. Evolution of organic matter during the mesophilic composting of lignocellulosic winery wastes. Journal of Environmental Management. 2013;**116**:18-26

[23] Adhikari BK, Barrington S, Martinez JKS. Effectiveness of three bulking agents for food waste composting. Waste Management. 2009;**29**:197-203

[24] Xi BD, Xi BD, He XS, Wei ZM, Jiang YH, Li MX, et al. Effect on inoculation methods on the composting efficiency of municipal solid wastes. Chemosphere. 2012;**88**: 744-750

[25] Iyengar SR, Bhave PP. In-vessel composting of household wastes. Waste Management. 2005;**26**:1070-1080

[26] Bernal MP, Alburquerque JA, Moral R. Composting of animal manures and chemical criteria for compost maturity assessment. A review. Biosource Technology. 2009;**100**: 5444-5453

[27] Senesia N, Plazab C, Brunettia G, Polob A. A comparative survey of recent results on humic-like fractions in organic amendments and effects on native soil humic substances. Soil Biology & Biochemistry. 2007;**39**:1244-1262

[28] Wange P, Changaa CM, Watsonb ME, Dickb WA, Chenc Y, Hoitinka HAJ. Maturity indices for composted dairy and pig manures. Soil Biology & Biochemistry. 2004;**25**: 767-776

[29] Provenzano MR, Oliveira SC, Silva MRS, Senesi N. Assessment of maturity degree of composts from domestic solid wastes by fluorescence and Fourier transform infrared spectroscopies. Journal of Agricultural and Food Chemistry. 2001;**49**:5874-5879

[30] Guo X, J.Gu, Gao H, Qin Q, Chen Z, Shao L, et al. Effects of Cu on metabolisms and enzyme activities of microbial communities in the process of composting. Biosource Technology. 2012;**108**:140-148

[31] Zittel R. Tratamento de Resíduos Orgânicos Domésticos, Tabaco de Cigarros Contrabandeados e Resíduos de Madeira em Biorreator. 2014. Dissertação (Mestrado em Química Aplicada) - Universidade Estadual de Ponta Grossa, Ponta Grossa, p.96, 2014.

[32] Associação Brasileira de Normas Técnicas —ABNT. NBR: 10007: Amostragem de Resíduos: Procedimentos [Online]. 2004 [cited 21 agosto 2013. Available from: www.iap.pr.gov.br/modules/conteudo/conteudo.php?conteudo=191

[33] Ministério da Agricultura, Pecuária e Abastecimento—MAPA. Instrução Normativa 25/2009 Art.1 Inc.III. [Online]. 07 dezembro 2009 [cited 2011. Available from: http://extranet.agricultura.gov.br/sislegis/action/detalhaAto.do?method=consultarLegislacaoFederal

[34] Kiehl EJ. Fetilizantes Orgânicos Piracicaba: Editora Agronomica "Ceres". 1985

[35] Hong Kong Organic Resource Centre—Hkorc. Compost and soil conditioner quality standards. Hong Kong Organic Resources Centre Tong K, editor. Hong Kong Baptist University. 2005

[36] Fialho LL, Silva WTL, Milori DMBP, Simões ML, Martin-Neto L. Characterization of organic matter from composting of different residues by physicochemical and spectroscopic methods. Bioresource Technology. 2010;**101**:1927-1934

[37] Juarez MF, Prähauser B, Walter A, Insam H, Franke-Whittle IH. Cocomposting of biowaste and wood ash, influence on a microbially driven-process. Waste Management. 2015;**46**: 155-164

[38] Mulec AO. Composting of the solid fraction of blackwater from a separation system with vacuum toilets—effects on the process and quality. Journal of Cleaner Production. 2016;**112**:4683-4690

[39] El Fels L, Hafidi M, Ouhdouch Y. Artemiasalina as a new index for assessment of acute cytotoxicity during co-composting of sewage sludge and lignocellulose waste. Waste Management. 2016;**50**:194-200

[40] Albrecht R, Le Petit J, Terrom G, Périssol C. Comparison between UV spectroscopic and Nirs to assess humification process during sewage sludge and green wastes co-composting. Bioresource Technology. 2011;**102**:4495-4500

[41] Spaccini R, Piccolo A. Molecular characteristics of humic acids extracted from compost as increasing maturity stages. Soil Biology & Biochemistry. 2009;**41**:1164-1172

[42] Yuan J, Chadwick D, Zhang D, Li G, Chen S, Luo W, et al. Effects of aeration rate on maturity and gaseous emissions during sewage sludge composting. Waste Management. 2016;**56**:403-410

[43] Zorpas AA, Inglezakis VJ, Loizidou M. Heavy metals fractionation before during and after composting of sewage sludge with natural zeolite. Waste Management. 2008;**28**:2054-2060

[44] Amir S, Jouraiphy A, Meddichb A, El Gharous M, Wintertonc P, Hafidi M. Structural study of humic acids during composting of activate sludge-green waste: Elemental analyses, FTIR and 13C NMR. Journal of Hazardous Materials. 2010;**177**:524-529

[45] Ali M, Bhatia A, Kazmi AA, Ahmed N. Characterization of high rate composting of vegetable market waste using Fourier transform-infrared (FT-IR) and thermal studies in three different seasons. Biodegradation. 2012;**23**:231-242

[46] El-Ouaqoudi FZ, El Fels L, Lemée L, Amblès A, Hafidi M. Evaluatin of lignocellulose compost stability and maturity using spectroscopic (FTIR) and thermal (TGA/TDA) analysis. Ecological Engineering. 2015;**75**:217-222

[47] Sarika D, Singh J, Prasad R, Vishan I, Varma VS, Kalamdhad ASS. Study of physicochemical and biochemical parameters during rotary drum composting of water hyacinth. International Journal Organic Waste Agriculture. 2014;**3**:63

[48] He WS, Xi BD, Wei ZM, Jiang YH, Geng CM, Yang Y. et al. Physicochemical and spectroscopic characteristics of dissolved organic matter extracted from municipal solid waste (MSW) and their influence on the landfill biological stability. Bioresource Technology. 2011;**102**:2322-2327

[49] Hachicha R, Hachicha S, Trabelsi I, Woodward S, Mechichi T. Evaluation of the fatty fraction during co-composting of olive oil industry wastes with animal manure: Maturity assessment of the product. Chemosphere. 2009;**75**:1383-1386

Chapter 7

Enhanced Anaerobic Digestion of Organic Waste

Abbass Jafari Kang and Qiuyan Yuan

Additional information is available at the end of the chapter

http://dx.doi.org/10.5772/intechopen.70148

Abstract

Anaerobic digestion (AD) of organic municipal solid waste (OMSW) is considered as a key element in sustainable municipal waste management due to its benefits for energy, environment, and economy. This process reduces emission of greenhouse gases, generates renewable natural gas, and produces fertilizers and soil amendments. Due to its advantages over other treatment methods and waste-to-energy technologies, anaerobic digestion has attracted more attention so that numerous research works in this area are performed. In this chapter, an overview of previous studies on anaerobic digestion using OMSW as the feedstock is presented. First, the principals of anaerobic digestion including chemical and biological pathways and microorganisms responsible for different steps of the process are discussed. Factors influencing the efficiency of the process such as temperature, pH, moisture content, retention time, organic loading rate and C/N ratio are also presented in this chapter. Different methods of pretreatment applied to enhance biogas production from anaerobic digestion of municipal solid waste are also discussed.

Keywords: anaerobic digestion, municipal solid waste, renewable resource, waste-to-energy, waste pretreatment, enhancing biogas production

1. Introduction

Municipal solid waste (MSW) management has become a serious environmental issue since the waste generation has rapidly increased with population explosion and economic development. Improper management of MSW can contribute to the degradation of environment quality [1]. For instance, the disposal of municipal solid waste in landfills can cause emission to the atmosphere as well as high nitrogen concentrations in the leachate [2, 3]. However, advancement of technology, establishment of environmental regulations, and emphasis on resource conservation and recovery have greatly reduced the environmental impacts of municipal solid waste management [4]. Emission of greenhouse gases through municipal solid waste management systems can be reduced by using a series of different treatment and disposal techniques such as sorting, aerobic composting, anaerobic digestion (AD), incineration, and landfill. Mata-Alvarez et al. [5] compared different municipal solid waste management systems including landfill (1.97 tons CO_2/ton MSW), incineration (1.67 CO_2/ton MSW), sorting-composting-landfill (1.61 CO_2/ton MSW), sorting-composting-incineration (1.41 CO_2/ton MSW), and sorting-anaerobic digestion-incineration-landfill (1.19 CO_2/ton MSW). The results showed that anaerobic digestion plays an important role in reducing CO_2 emission from municipal solid waste.

Bioprocessing of organic fraction of municipal solid waste that comprises composting and anaerobic digestion is considered as a viable means of transforming organic wastes into products that can be used safely and beneficially as biofertilizers and soil conditioners [6]. Aerobic treatment has been found to cause large and uncontrolled emission of volatile compounds such as ketones, aldehydes, and ammonia [5]. Additionally, composting is an energy consuming process (around 30–35 kWh/ton waste), while anaerobic digestion is a net energy producing process (100–150 kWh/ton waste) [7].

Energy from waste is seen as one of the most dominant future renewable energy sources, especially since that a continuous power generation from these sources can be guaranteed [8]. The most important property of alternative energy source is their environmental compatibility [9]. Various methods have been applied to convert waste to energy such as combustion, gasification, pyrolysis, fermentation, and anaerobic digestion. Among these methods, anaerobic digestion has attracted more attention because of following advantages: anaerobic digestion can process a variety of biomass materials (sewage sludge, municipal solid waste, agricultural wastes, manure, and industrial wastes); this process can easily treat wet wastes, which are problematic in other methods such as combustion; anaerobic digestion obtains valuable products which are useful for soil fertilization and energy generation; compared to common waste management processes such as incineration, pyrolysis, and gasification, this process causes the least amount of air and solid pollution. The other advantage is the small size of AD plants, which offers less footprint [10–12].

In this chapter, an overview of anaerobic digestion of municipal solid waste including fundamental of anaerobic digestion, microbiology of the process, important operating factors, and the techniques used for enhancing anaerobic digestion of municipal solid waste is presented.

2. Principals of anaerobic digestion

Anaerobic digestion is a series of biological processes in which microorganisms break down biodegradable material in the absence of oxygen. As shown in **Figure 1**, this process occurs in four stages including hydrolysis, acidogenesis, acetogenesis, and methanogenesis.

Hydrolysis, as the first stage of anaerobic digestion, is conversion of insoluble complex organic matter (carbohydrates, proteins, and lipids) into soluble molecules (sugars, amino acids, and long chain fatty acids). Hydrolysis reactions are carried out by extracellular enzymes called

http://dx.doi.org/10.5772/intechopen.70148

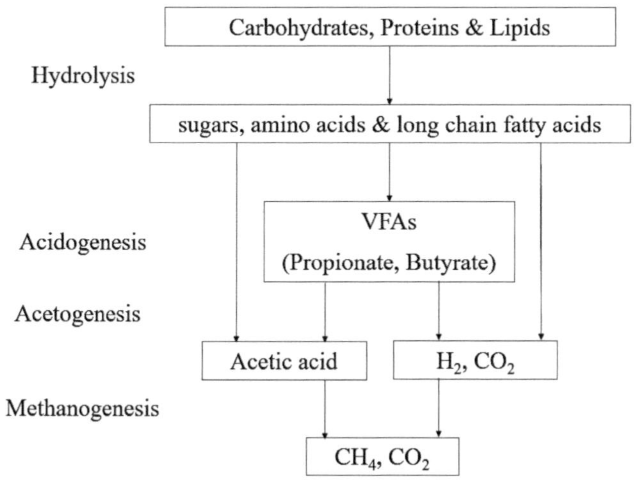

Figure 1. Process flow of the degradation of organic material through anaerobic digestion [16].

hydrolase. These hydrolases can be esterase (enzymes that hydrolyze ester bonds in lipids), glycosidases (enzymes that hydrolyze glycosidic bonds in carbohydrates), and peptidases (enzymes that hydrolyze peptide bonds in proteins) [13]. These enzymes are produced by microorganisms called hydrolytic bacteria. *Clostridium, Proteus vulgaris, peptococcus, bacteroides, bacillus, vibrio, acetivibrio cellulolyiticus, staphylococcus,* and *micrococcus* have been reported as typical species of hydrolytic bacteria [14]. For example, *Cellulomonas* bacterium produces cellulase enzyme which can degrade polysaccharides into simple sugar; *Bacillus* bacterium converts proteins to amino acids by producing protease enzyme; and lipase enzyme produced by *Mycobacterium*, converts lipids into fatty acids [15]. Hydrolysis was reported [13] as the rate-limiting step in anaerobic digestion process due to the slow depolymerization of the insoluble complex organic matter by hydrolytic bacteria.

In acidogenesis stage, fermentative bacteria convert soluble molecules produced in the hydrolysis stage into volatile fatty acids (propionate, b), lactate, alcohols, and carbon dioxide. There are different fermentation pathways each of which is carried out by different bacterial species. Some genera of bacteria, which carry out fermentation pathways in anaerobic digestion, are as follows: *Saccharomyces* (alcohol fermentation), *Butyribacterium* and *Clostridium* (butyrate fermentation), *Lactobacillus* and *Streptococcus* (lactate fermentation), and *Clostridium* (propionate fermentation). Acetate is also produced in this step by a group of bacteria called acetate-forming fermentative bacteria. *Acetobacterium, Clostridium, Eubacterium,* and *Sporomusa* are typical species of acetate-forming fermentative bacteria [14, 16].

In the third stage of anaerobic digestion, acetogenic bacteria transform volatile fatty acids (VFAs) and alcohols into acetate, H_2 and CO_2. Different species have been identified as acetogens. *Syntrophobacter wolinii* and *Smithella propionica* are identified bacteria which form acetate by consuming butyrate and propionate, respectively. Some other species such as *Syntrophobacter fumaroxidans, Syntrophomonas wolfei, Pelotomaculum thermopropionicum,* and *Pelotomaculum schinkii* have been identified as acetogenic bacteria which convert VFAs to formate, H_2 and CO_2. Clostridium aceticum is another identified microorganism which produces acetate from H_2 and CO_2 [14].

Methanogenesis is the final stage of anaerobic digestion in which formation of methane gas from acetate and molecular hydrogen occurs. Methanogens play a vital role as the consumer of acetogenesis products due to the fact that accumulation of hydrogen produced in acetogenesis can terminate activity of acetate-forming bacteria [15]. Different species have been identified as methanogenic bacteria including: (i) species which convert acetate to methane and carbon dioxide (acetoclastic methanogenic pathway) such as *Methanothrix soehngenii* and *Methanosaeta concilii*; (ii) species which produce methane from H_2 and CO_2 (hydrogenotrophic methanogenic pathway) such as *Methanobacterium bryantii, Methanobacterium thermoautotrophicum*, and *Methanobrevibacter arboriphilus*; and (iii) species which consume formate, hydrogen, and carbon dioxide and produce methane such as *Methanobacterium formicicum, Methanobrevibacter smithii*, and *Methanococcus voltae* [14].

3. Important operating factors

The anaerobic digestion of organic material is a complex process, involving a number of different degradation steps. The microorganisms that participate in the process may be specific for each degradation step and thus could have different environmental requirements [17]. Parameters affecting anaerobic digestion include temperature, moisture content, retention time, pH, organic loading rate, and C/N ratio.

3.1. Temperature

Operating temperature is the most important factor determining the performances of anaerobic digestion because it is an essential condition for the survival and growth of the microorganisms [18]. It also determines the values of the main kinetic parameters for the process and, hence, the rate of the microbiological process. There are two range of temperature with maximum anaerobic digestion rate (gas production rates, bacteria growth rate, and substrate degradation rate): thermophilic (50–60°C) and mesophilic (30–40°C) [19].

Mesophilic and thermophilic anaerobic digestion have been widely used for biogas production from various types of waste and the results have shown these processes to have different advantages and disadvantages as listed in **Table 1**.

Results of a comparative study [21] on anaerobic digestion of organic municipal solid waste (OMSW) under mesophilic (35°C) ,and thermophilic (50°C) conditions showed that microbial activity is favored working at thermophilic range; hence, higher specific growth rate and methane yields were achieved in the thermophilic anaerobic digestion. Thermophilic digesters presented a higher rate of soluble chemical oxygen demand (sCOD) removal and methane production rate compared with mesophilic digesters in a study on anaerobic digestion of food waste [22]. Values reported for microbial activity (maximum specific growth rate) and methane production (specific methane yield) of anaerobic digestion of the organic fraction of municipal solid waste are presented in **Table 2**.

	Advantages	Disadvantages
Thermophilic	Higher loading rate Higher methane production. Higher temperature shortens the required retention time. higher pathogen destruction	Very sensitive to toxins and small environmental changes The process is less stable, as the microbial population is less divers. The system is harder to maintain Additional energy input is required for heating
Mesophilic	Operates with robust microorganisms which tolerate greater changes in the environment. The system is more stable and easier to maintain. Smaller energy expense	Longer retention time Lower biogas production

Table 1. Comparison of thermophilic and mesophilic anaerobic digestion [10, 17, 18, 20].

3.2. Moisture content

Moisture content is one of the most important factors affecting anaerobic digestion. Moisture was reported [10] to aid digestion by (i) controlling cell turgidity; (ii) transporting nutrients, intermediates, products, enzymes and microorganisms; (iii) reacting in hydrolysis of complex organic matters; and (iv) modifying the shape of enzymes and other macromolecules [17]. High

Bacterial group	Mesophilic		Thermophilic		Refs.
	μ_{max} (d^{-1})	CH_4 production (m^3 CH_4/g VS)	μ_{max} (d^{-1})	CH_4 production (m^3 CH_4/g VS)	
Acetoclastic methanogens	0.192–0.256	0.0079	0.243–0.410	0.0149	[21]
Hydrolysis	0.024	0.0019			[23]
Acidogens	2.4				
Acetoclastic methanogens	0.1392				
Hydrogenotrophic methanogens	1.39				
Hydrolysis			0.08–0.18	0.016	[24]
Acidogens			0.13–0.16		
Acetoclastic methanogens			0.23–0.28		
Hydrogenotrophic methanogens			0.33–0.40		
Overall			0.118–0.178	0.023 m^3 CH_4/g COD	[25]
Acetoclastic methanogens	0.13–0.19	0.0047–0.0079			[26]
All methanogens	0.15–0.26				
Overall			0.58	–	[27]

Table 2. Microbial activity and methane production of mesophilic and thermophilic anaerobic digestion of OMSW.

moisture contents usually facilitate the anaerobic digestion due to the fact that water contents are likely to affect the process performance by dissolving readily degradable organic matter.

Based on the total solids content of the slurry in the digester reactor, anaerobic digestion processes are classified to low solids or wet digestion (less than 10% TS), medium solids or semidry (10–20% TS) and high solids (more than 20% TS). Most of the studies on degradation of organic fraction of municipal solid waste were performed using dry anaerobic digestion process due to the high-solid content of OMSW [21, 24, 27, 28]. However, adding water or codigesting with low-solid wastes such as sewage sludge and manure can increase the moisture content of OMSW and make it suitable for semidry anaerobic digestion process [26, 29].

Lay et al. [30] reported that increasing initial moisture content of mesophilic anaerobic digesters from 90 to 96% increased the methanogenic activity in high-solids sludge digestion. In another study [28], digesters operated at higher initial moisture content obtained higher methane production rate, as well as better dissolved organic carbon (DOC) removal efficiency in mesophilic anaerobic digestion of OMSW. However, it was reported [31] that increasing the moisture content of OMSW decreased methane production rate of anaerobic digesters with periodic cycles of leachate drainage and water addition. The bioreactors operating at 80% moisture content presented a poorer volatile solids (VS) content compared to the ones operating at 70% moisture content due to the fact that water readditions into the bioreactors could contribute to washing out of nutrients and microorganisms.

3.3. Retention time

Retention time is the required time for decomposition of organic matter, which is determined by measuring chemical oxygen demand (COD) or biochemical oxygen demand (BOD) of the influent and the effluent. Longer retention time will result in more degradation of organic matter [18]. Required retention time for complete AD is controlled by the applied technology, process temperature, and waste composition. Mesophilic anaerobic digestion requires retention time of 8–40 days; while less thermophilic AD offers less retention time [32]. Fdez-Güelfo et al. [27] investigated the effect of solids retention time (SRT), from 8 to 40 days, on the dry thermophilic anaerobic digestion of OMSW. They reported that SRT of 15 days obtained the highest VS removal and methane yield.

Reducing retention time reduces the volume required for the reactor and consequently reduces the capital costs of anaerobic digestion. Therefore, different approaches have been suggested [20] for reducing the retention time such as mixing, decreasing solid content, separating stages of anaerobic digestion, and alternating flow pattern and pretreatment.

Proper mixing ensures that bacteria have rapid access to as many digestible surfaces as possible and that environmental characteristics are consistent throughout the digester [20]. Recirculating water and biogas in the chamber to keep material moving has been used as a promising mixing method to enhance anaerobic digestion. Cavianto et al. [33] reported that water recirculation improved methane production in a two-phase thermophilic anaerobic process with hydraulic retention time of 3 days. Decreasing solid content can reduce the retention time, because bacteria can more easily access liquid substrate and because the relevant reactions require water. Additionally, mixing is more complete when the solid content is lower [20]. Retention time can be reduced by separating the stages of the digestion into individual chambers so that the bacterial population in each chamber is optimized for its purpose [20]. Two-phase anaerobic digestion has been used advantageously to treat wastes with high-solid content such as municipal solid waste and reported to be warranted from the kinetic point of view [34].

Using various methods of pretreating waste (discussed later) can also reduce the retention time by increasing digestibility.

3.4. pH

Operating pH is another important factor due to the sensitivity of methanogenic bacteria, their growth as well as methane production to acidic. Biological activities during different stages of anaerobic digestion change the pH level. Production of organic acids during the acetogenesis phase lower the pH down to 5 which is lethal for methanogens and can cause digester failure [18].

High VFA levels could occur due to overloading, poor mixing, nutrient shortage, variation of temperature, and loss of bacteria in the discharge. If there is enough alkalinity available, the acids may be buffered; thus, buffering reagents may be needed. However, buffering can be provided by the reaction of the ammonium ions with bicarbonate ions to form ammonium bicarbonate [10].

In the start-up when fresh waste is introduced, before methanogenesis stage starts, organic acids are formed. This lowers the pH. Therefore, pH control is delicate during the early stages. Addition of buffers such as calcium carbonate or lime to the system in order to increase the pH is necessary [20].

Ward et al. [11] reported that the pH range of 6.8–7.2 is ideal for anaerobic digestion. They also reported that the optimal pH of methanogenesis is around pH 7.0 and the optimum pH of hydrolysis and acidogenesis is between pH 5.5 and 6.5. This is an important reason why some researchers prefer the separation of the hydrolysis/acidification and acetogenesis/methanogenesis processes in two-stage processes. Recirculation of process liquid was also reported to have a beneficial effect on the performance of anaerobic codigestion of OMSW and manure by stabilization of the pH [29].

Zhang et al. [35] investigated the effect of pH, the first stage (hydrolysis and acidogenesis) of anaerobic digestion of kitchen waste adjusting pH values of 5, 7, 9, and 11. They reported that pH adjustment improved both hydrolysis and acidogenesis rates as well as TS removal rate and biogas production during the two-phase anaerobic digestion of kitchen wastes.

3.5. Organic loading rate

Organic loading rate (OLR) is another important parameter to control since the biogas production rate is highly dependent on the loading rate [36]. Basically, for the higher volatile solids, more bacteria is needed for anaerobic digestion. Increasing the OLR increases the population of acidogenic bacteria which produce acids and multiply rapidly. However, methanogenic bacteria that take longer to increase their populations would not be able to consume the acids at the same pace [20]. Consequently, pH of the system will fall, which can kill methanogenic bacteria and lead to the crash of the system [18].

Dhar et al. [48] investigated the effect of organic loading rate during anaerobic digestion of municipal solid waste using mesophilic reactors with initial loading of 5.1 and 10.4 g/L sCOD. The results showed that the reactor with higher organic loading rate obtained a higher methane yield (168 mL $CH_4/gVS_{removed}$) as well as higher CODs reduction (84.2%) than the other reactor (methane yield of 101 mL $CH_4/gVS_{removed}$ and 78% CODs reduction). Bouallagui et al. [37] tested anaerobic digestion of fruit and vegetable wastes using three two-phase mesophilic digesters operated with organic loading rates of 3.7, 7.5, and 10.1 g COD/L.d. The results indicated that with the increase in the organic loading rated, the biogas yield increased from 363 to 448 L/kg COD_{input} and COD removal of the total process increased from 79 to 96%.

3.6. Carbon and nitrogen content

As a matter of fact, carbon constitutes the energy source for the microorganisms and nitrogen serves to enhance microbial growth. If the amount of nitrogen is limiting, microbial populations will remain small and it will take longer to decompose the available carbon [38]. Excess nitrogen, on the other hand, inhibits the anaerobic digestion process. Since it has been found that microorganisms utilize carbon 25–30 times faster than nitrogen, a ratio of 20–30:1 was reported as the optimum carbon/nitrogen ratio for anaerobic digestion [11]. Elsewhere, a nutrient ratio of the elements C:N:P:S (carbon:nitrogen:phosphorous:sulfur) at 600:15:5:3 was reported sufficient for methanization [17]. A low C/N ratio, or too much nitrogen, can cause ammonia to accumulate which would lead to pH values above 8.5 [20]. In order to improve the nutrition and C/N ratios, codigestion of different organic mixtures has been employed. C/N ratio and methane yield reported in some of the studies on codigestion of municipal solid waste and other types of organic waste are summarized in **Table 3**.

Sosnowski et al. [39] investigated anaerobic codigestion of sewage sludge (primary sludge and thickened excess activated sludge) and organic fraction of municipal solid wastes (25% total volume). The results showed that addition of the OMSW to the sewage sludge improved the C/N ratio from 9/1 to 14/1 and increased cumulative biogas produced.

Heo et al. [40] studied anaerobic biodegradability of food waste (FW), waste activated sludge (WAS) in a single-stage anaerobic digester operating at 35°C. They reported that as the FW proportion of the mixture increased from 10 to 90%, C/N ratio of the mixtures improved (from 6 to 15), biodegradation of the mixture increased and the methane production increased.

In another study [41], the mesophilic anaerobic codigestion of food waste and cattle manure was tested. The results indicated that the total methane production was enhanced in codigestion, with an optimum food waste (FM) to cattle manure (CM) ratio of 2:1. The C/N ratio and the higher biodegradation of lipids were the main reasons for the biogas production improvement.

AD process	substrates	Dry weight ratio	C/N	Methane yield (LCH_4/gVS)	Refs.
Thermophilic batch	OMSW:Sludge	2:1	14.19	0.14	[39]
Two-stage thermophilic-mesophilic	OMSW:Sludge	2:1	14.19	0.18	[39]
Single-stage stage mesophilic	Food waste:Sludge	1:9	5:97	0.186	[40]
		3:7	6.99	0.215	
		1:1	8.9	0.321	
		7:3	11	0.336	
		9:1	14.7	0.346	
Mesophilic batch	Foodwaste:Cattle manure	2:1	15.8	0.388	[41]
Two-phase	OMSW:Cow manure	10:1	20	0.10	[42]
Mesophilic batch	OMSW:Sludge	1:34	17.68	0.15	[43]
		1:19	20.55	0.20	
Singles stage	Food waste:Sludge	1:2.4	7.1	0.303	[44]
		1:0.9	10.2	0.350	
		1:0.4	11.4	0.400	
Mesophilic	OMSW		14.1	0.382	[45]
	OMSW:Vegetable oil	5:1		0.699	
	OMSW:Animal fat	5:1		0.508	
	OMSW:Cellulose	5:1		0.254	
	OMSW:Protein	5:1		0.288	

Table 3. Results of some studies on codigestion of municipal solid waste and other types of waste.

Nitrogen plays an important role in anaerobic digestion due to the fact that in the form of ammonium, nitrogen contributes to the stabilization of the pH value in the reactor. Nitrogen can cause problems in anaerobic digestion because of its metabolic products (ammonia/ammonium) [46]. Ammonium ion may inhibit the methane producing enzymes directly; while ammonia molecule may diffuse into bacterial cells, which causes intracellular pH change by conversion into ammonium and consequently, inhibition of specific enzyme reactions [47]. The NH_3 fraction of total ammonia nitrogen depends on pH and temperature. For three different operating temperatures, the dissociation balance of ammonia and ammonium with change in pH is plotted in **Figure 2**, showing that at high value of pH rapid conversion of ionized

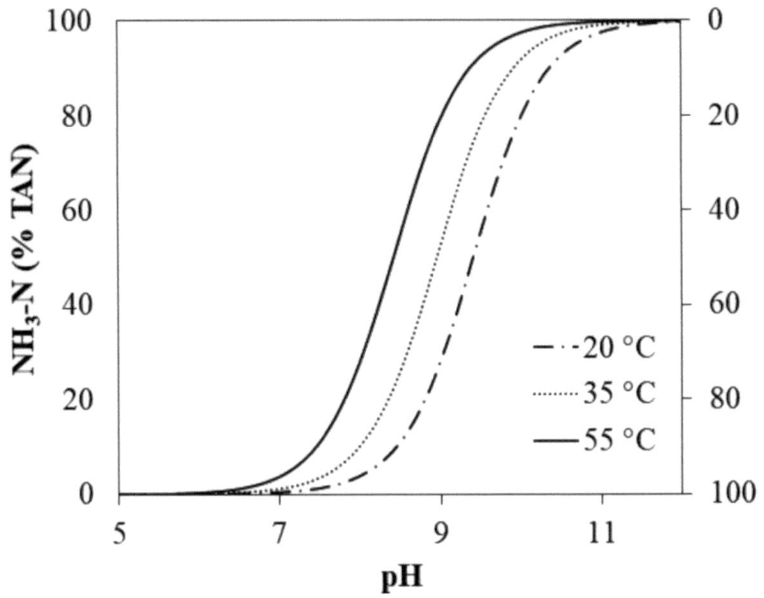

Figure 2. Dissociation balance between ammonia and ammonium at different operating temperatures, adapted from Ref. [46].

ammonia nitrogen (NH_4^+) into free ammonia nitrogen (NH_3) occurs. Increasing amount of NH_3 inhibits the methanogenic microflora and resulted in accumulation of VFAs, which again leads to decrease in pH and thereby declining the concentration of NH_3. The interaction between NH_3, VFAs and pH may lead to lower methane yield [46]. Due to the effect of temperature on dissociation of ammonia/ammonium, anaerobic digestion can be more easily inhibited and less stable at thermophilic temperature than at mesophilic temperature [47].

The ammonia-induced inhibition was reported to occur during the anaerobic digestion of organic waste materials rich in proteins. The inhibiting concentrations was found between 30 and 100 mg/L ammonia or 4000 and 6000 mg/L ammonia (at pH value ≤7 and temperature ≤30°C) [46].

Different strategies such as pH and temperature control, acclimation of microflora, and diluting reactor content were suggested in order to prevail over the ammonia inhibition during the anaerobic digestion process [47].

4. Municipal solid waste as the feedstock

Municipal solid waste is the most variable feedstock as the methane yield value depends not only on the sorting method, but also on the location from which the material was sourced and the time of year of collection [11]. Anaerobic digestion became possible because of the introduction of source separation collection of a clean biodegradable fraction, otherwise a presorting step is necessary to remove compounds which are not suitable for anaerobic digestion. However, adding a presorting step significantly increases the treatment costs [8].

Food waste is a significant proportion of organic fraction of residential waste and contains a high moisture content, which can generate leachate and odor. Other contents of OMSW are yard waste and paper products. The characteristics of municipal solid waste used as the feedstock in anaerobic digestion process along with the reactor type, operation condition, VS or COD removal efficiency and methane yield reported by some authors are presented in **Table 4**.

Reactor type	Vol. (L)	T (°C)	Waste characteristics				pH	HRT (d)	OLR gCOD /L.d	VS or COD removal	CH_4 yield (L/gVS)	Refs.
			TS (g/L)	VS (g/L)	COD (g/L)	TKN (g/L)						
Single stage	2	38	24.7 g/kg	78.6 % TS	21.3	–	6.4–7.5	4	3.85	84% COD_r	0.168	[48]
Two phase	1.5, 5	35	100	88	120	3.8	6.9–7.5	10	1.65	96% COD_r	0.450 (L/gCOD)	[37]
Single stage	4.8	37	184	172	176	3.1	6.5–7.5	19	9.65	64% VS_r	5.3 (L/L_Rd)	[49]
		55								65% VS_r	5.6 (L/L_Rd)	
Single stage	1	37	29%	77 % TS	–	1.83	–	21	2.36	–	0.382	[45]
Single stage	1	37	1.56%	54.14 % TS	–	3.03 % TS	>6.5	35	–	63% VS_r	0.17	[43]
Two phase	200, 760	55	241 g/kg	203 g/kg	206 g/kg	7.2	4.3, 7.6	6	21 g VS/Ld	–	0.78	[33]
Single phase	3000	55	201.4 g/kg	124.3 g/kg	85.9 g/kg	14 g/kg	7.1–7.7	13.5	9.2 g VS/Ld	–	0.23	[49]
Single phase	4.5	55	0.90 g/g	0.71 g/g	–	–	7.7	15	11.8 g VS/Ld	89% VS_r	1.15 L/L_Rd	[27]
Single phase	200	38	67%	75 % TS	1100	11.7	7.2	20	–	43.2% COD_r	0.19	[23]
Single phase	1	35	18.5%	17%	–	3.16%	7.3–7.5	18	8 gVS/ Ld	–	0.41	[41]
Two stage	9 dm^3 14 dm^3	56 36	7%	–	–	–	–	8.9 20.9	2.76 gVS / dm^3d	–	0.3	[39]

Table 4. Results of some studies on anaerobic digestion OMSW.

5. Enhancement of anaerobic digestion

In order to enhance biogas production and volatile solids reduction, various pretreatment techniques have been applied. These techniques can be classified as physical pretreatment, chemical pretreatment, biological pretreatment, and thermal pretreatment and combination of these methods such as thermochemical pretreatment.

5.1. Physical pretreatment

Particle size reduction, also known as mechanical pretreatment, is a classic method to increase the efficiency of anaerobic digestion process. This type of treatment improves the biological process by increasing the specific surface available to the microorganisms and leads to more rapid digestion [36]. In addition, the size of the feedstock should not be too large otherwise it would result in the clogging of the digester [5]. Zhang et al. [41] reported that change in the particle size of OMSW did not change the methane yield in the wet anaerobic digestion system. They also reported that the finer materials in the dry digesters caused process failure at high organic loading rates due to the rapid acidification. Advantages of mechanical pretreatment include no odor generation, an easy implementation, better dewaterability of the final anaerobic residue and moderate energy consumption [50]. Nopharatana et al. [23] used hammer-milled MSW with an average particle size of 2 mm and coarsely shredded MSW with an average particle size of 5 cm as feedstock in mesophilic anaerobic digestion. The results showed that reduction of particle size increased the methane yield. However, the author reported that for particle sizes considered, the surface area has no appreciable effect on the kinetics of digestion. Elsewhere [51], methane production rate of mesophilic anaerobic digestion process increased 28% when the mean particle size of food waste was decreased from 0.888 to 0.718 mm by bead milling pretreatment. The authors also reported that excessive size reduction of the substrate decreased methane production due to VFA accumulation.

Rotary drum reactor process was used as a mechanical pretreatment method providing an effective means for separating the organic fraction of municipal solid waste prior to anaerobic digestion [52]. A methane yield of 0.522 m^3 CH_4/kgVS was achieved from thermophilic anaerobic digestion of municipal solid waste pretreated in a rotary drum reactor for 1 day. However, lower methane yields were reported for the same MSW pretreated with rotary drum for longer retention times; 0.509 and 0.489 m^3 CH_4/kgVS for 2–3 days retention time was reported, respectively.

Hansen et al. [53] used different pretreatment technologies such as screw press, disc screen and shredder plus magnet prior the anaerobic digestion process. Considering the sorting effect of screw press and disc screen, the screw press was reported to have a larger selective effect than the disc screen by routing more water and easy degradable organic matter and less slowly degradable organic matter. In terms of biogas production, anaerobic digestion of MSW pretreated with the shredder plus magnet yielded a higher amount of methane (102 m^3 CH_4/ton waste) than those of the two other pretreatment methods (40–60 m^3 CH_4/ton waste).

5.2. Chemical pretreatment

Chemical pretreatment methods including acidic pretreatment, alkali pretreatment and ozonation are used to achieve the destruction of the organic compounds and consequently enhance the biogas production and improve the hydrolysis rate [50]. Alkali treatment was reported to be particularly advantageous when using plant materials with high lignin content in anaerobic [11]. Torres et al. [54] investigated the effect of alkali pretreatment on anaerobic digestion of OMSW by lime addition ($Ca(OH)_2$). The results indicated that the alkaline pretreatment improved anaerobic digestion by increasing the soluble COD (enhancing the COD solubilization). Consequently, the methane yield increase from 0.055 to 0.15 m^3 CH_4/kgVS.

Acid pretreatment was reported to be more desirable for lignocellulose substrates, not only because it breaks down the lignin, but also because the hydrolytic microbes are capable of acclimating to acidic conditions [11]. However, this type of pretreatment has some disadvantages. Strong acidic pretreatment may result in the production of inhibitory by-products such as furfural and hydroxymethylfurfural. Other disadvantages associated with the acid pretreatment include the loss of fermentable sugar due to the increased degradation of complex substrates, high cost of acids and the additional cost for neutralizing the acidic conditions prior to the anaerobic digestion [50].

Ozone is a strong oxidant, decomposes itself into radicals and reacts with organic substrates. As the result, the recalcitrant compounds become more biodegradable and accessible to the anaerobic [50]. Cesaro et al. [55] investigated the effect of ozone pretreatment (at different doses) on mesophilic anaerobic digestion of organic fraction of municipal solid waste. The results indicated that ozonation with a dose of 0.16 gO_3/gTS increased biogas volume by 37%.

Ultrasonic pretreatment is another technique which is commonly used to break down complex polymers in the treatment of sewage sludge. Mechanical shear forces caused by ultrasonic pretreatment as a key factor for sludge disintegration can significantly alter the sludge characteristics in sewage sludge treatment and increase the methane production [11]. Ultrasonic pretreatment of OMSW obtained 16% increase in biogas production of mesophilic anaerobic digestion [55].

5.3. Biological pretreatment

Aerobic and anaerobic methods can be used prior to anaerobic digestion to enhance the biogas production as well as VS reduction. As an anaerobic pretreatment, the first step (hydrolytic-acidogenic) of a two-phase AD process acts as a biological pretreatment method. The advantages of such systems include: (i) increased stability with better pH control; (ii) higher loading rate; (iii) increased specific activity of methanogens resulting in a higher methane yield; (iv) increased VS reduction and (v) high potential for removing pathogens [50]. The addition of microbial strains (such as cellulolytic bacteria and fungi or cell lysate) increases the substrate digestibility [56]. Strains of some bacteria and fungi have also been found to enhance gas production by stimulating the activity of particular enzymes involved in cellulose degradation [36]. Preaeration was found to improve the thermophilic anaerobic digestion of OMSW by reducing the excess easily degradable organic compounds which are the main cause of acidification during the start-up. Charles et al. [57] reported that preaeration of OMSW for 48 hours generated enough biological heat to increase the temperature of bulk OMSW to 60°C which was sufficient self-heating of the bulk OMSW for the start-up of thermophilic anaerobic digestion without the need for an external heat source. Fdez-Güelfo et al. [25] investigated the effect of different biological pretreatments such as using mature compost, sludge and the fungus *Aspergillus awamori* on the anaerobic digestion of OMSW. The results showed that pretreatment with mature compost obtaining the highest increases in DOC (dissolved organic carbon) removal and methane production was the best of all the pretreatments. Sosnowski et al. [39] investigated the anaerobic codigestion of OMSW and sludge in thermophilic batch wise and two-stage quasi-continuous, acidogenic digestion under thermophilic conditions (56.8°C) and mesophilic methane fermentation (36.8°C). They reported that the separation of acidogenic and methanogenic stages two-stage anaerobic digestion was effective increased the methane yield from 0.14 to 0.18 L CH_4/gVS. Elsewhere [37], phase separation with conventional anaerobic sequencing batch reactors resulted in high process stability, significant biogas productivity and better effluent quality from fruit and vegetable wastes anaerobic digestion.

5.4. Thermal pretreatment

The use of high temperatures can also be used as a pretreatment method. The main effect of thermal pretreatment is the disintegration of cell membranes, thus resulting in solubilization of organic compounds. Thermal pretreatment also leads to pathogen removal, improves dewatering performance and reduces viscosity of the digestate, with subsequent enhancement of digestate handling [50]. Results of a study [58] on the effect of thermal pretreatment of sludge, kitchen waste and fruit/vegetable waste showed that thermal pretreatment at 175°C obtained a doubled methane production rate. While, the thermal pretreatment decreased the methane production by 8 and 12% for kitchen waste and fruit/vegetable waste, respectively. Shahriari et al. [59] used microwave heating at different temperature ranging from 115 to 175°C to enhance anaerobic digestion of OMSW. The biogas production increased 4–7% using 115 and 145°C pretreatment while pretreatment at 175°C decreased the biogas production due to formation of refractory compounds, inhibiting the digestion.

6. Anaerobic digestion for rural communities

Need for fertilizers and soil conditioners in rural areas and also, popularity of biofertilizer compared to chemical products make aerobic/anaerobic composting favorable [60, 61]. Due the aforementioned advantages of anaerobic digestion such as minimizing greenhouse gas emissions, reducing pathogens and sustainable energy production, this method can be considered as the best option for the treatment of organic solid waste in urban as well as rural areas.

In addition to organic fraction of household solid waste generated in rural area, there are other sources including plantation solid waste, residues from animal feed production and livestock manure [62, 63]. This offers the possibility of codigestion which is beneficial for the enhancement of biogas production by adjusting C/N ratio and moisture content mentioned before. Furthermore, different types of "energy crops" (such as maize, grass, and cereals) which have become attractive for biomethanation can be used as cosubstrate in anaerobic digestion and increase the methane yield [63, 64].

Anaerobic digestion in rural communities can be carried out by small-scale or family size biogas plants which utilize animal manure [65, 66]. Rural household biogas was reported to promote agricultural structural adjustment, raise rural incomes, enhance the ecology of rural areas, and improve the quality of both rural life and agricultural products [67]. The other opportunity for anaerobic digestion in rural areas is central anaerobic digestion (CAD) in which different farms cooperate to feed a single large digestion plant with a variety of cosubstrates. CAD plants have the potential to optimize the biogas production with codigestion and benefit a large community [68].

Small-scale household digesters are mostly built below ground and the produced biogas is mainly used for cooking. They are most commonly used in China and India [66]. Two types of digesters are used for household biogas production: constructed digesters (set up in 1920s) which are made of clay, brick, and concrete and commercial biogas digesters (introduced in 2000) made of glass fiber-reinforced plastics (GRP) [67].

7. Summary

Among the waste treatment technologies, anaerobic digestion can be considered as the best available technique for the treatment of organic fraction of municipal waste treatment technologies. This technology offers benefits for environment, energy, and economy.

This biological process consists of different stages. Through these stages, organic fraction of municipal solid waste is converted to biogas by different pathways in each of which, different species of microorganisms are responsible. Environmental factors such as temperature, moisture content, pH, organic loading rate, and carbon/nitrogen ratio, which influence different stages of the process and consequently the efficiency of the whole system. One of the important factor is C/N ratio of the feedstock which has been suggested to be in the range of 25–30 to obtain the best efficiency. In order to improve the nutrition and C/N ratios, codigestion of OMSW and other organic wastes can be employed.

Availability of different types of waste materials (which can be codigested) in rural areas as well as need for biofertilizer makes anaerobic digestion attractive in these areas. Biogas production can be carried out using household digesters or in a central anaerobic digestion plant which benefits a large community.

Author details

Abbass Jafari Kang and Qiuyan Yuan*

*Address all correspondence to: qiuyan.yuan@umanitoba.ca

Department of Civil Engineering, University of Manitoba, Winnipeg, Manitoba, Canada

References

[1] Behzad N, Ahmad R, Saied P, Elmira S, Mokhtar M. Challenges of solid waste management in Malaysia. Research Journal of Chemistry and Environment. 2011;**15**(2):597-600

[2] Jokela JP, Rintala JA. Anaerobic solubilisation of nitrogen from municipal solid waste (MSW). Reviews in Environmental Science and Biotechnology. 2003 Mar 1;**2**(1):67-77

[3] Nguyen PH, Kuruparan P, Visvanathan C. Anaerobic digestion of municipal solid waste as a treatment prior to landfill. Bioresource Technology. 2007 Jan 31;**98**(2):380-387

[4] Weitz KA, Thorneloe SA, Nishtala SR, Yarkosky S, Zannes M. The impact of municipal solid waste management on greenhouse gas emissions in the United States. Journal of the Air & Waste Management Association. 2002 Sep 1;**52**(9):1000-1011

[5] Mata-Alvarez J, Mace S, Llabres P. Anaerobic digestion of organic solid wastes. An overview of research achievements and perspectives. Bioresource Technology. 2000 Aug 31;**74**(1):3-16

[6] Kumar S. Composting of municipal solid waste. Critical Reviews in Biotechnology. 2011 Jun 1;**31**(2):112-136

[7] Braber K. Anaerobic digestion of municipal solid waste: A modern waste disposal option on the verge of breakthrough. Biomass and Bioenergy. 1995 Jan 1;**9**(1-5):365-376

[8] Appels L, Lauwers J, Degrève J, Helsen L, Lievens B, Willems K, Van Impe J, Dewil R. Anaerobic digestion in global bio-energy production: Potential and research challenges. Renewable and Sustainable Energy Reviews. 2011 Dec 31;**15**(9):4295-4301

[9] Kothari R, Tyagi VV, Pathak A. Waste-to-energy: A way from renewable energy sources to sustainable development. Renewable and Sustainable Energy Reviews. 2010 Dec 31;**14**(9):3164-3170

[10] Potts LG, Martin DJ. Anaerobic digestion, gasification, and pyrolysis. in Encyclopedia of Life Support Systems: Waste Management and Minimization. 2009 Sep 17:194-294

[11] Ward AJ, Hobbs PJ, Holliman PJ, Jones DL. Optimisation of the anaerobic digestion of agricultural resources. Bioresource Technology. 2008 Nov 30;**99**(17):7928-7940

[12] Chynoweth DP, Owens JM, Legrand R. Renewable methane from anaerobic digestion of biomass. Renewable Energy. 2001 Mar 31;**22**(1):1-8

[13] Palmisano AC, Barlaz MA. Microbiology of Solid Waste. Boca Raton, FL: CRC Press; 1996 Aug 16

[14] Amani T, Nosrati M, Sreekrishnan TR. Anaerobic digestion from the viewpoint of microbiological, chemical, and operational aspects—a review. Environmental Reviews. 2010 Aug 12;**18**(NA):255-278

[15] Gerardi MH. The Microbiology of Anaerobic Digesters. Hoboken, NJ: John Wiley & Sons; 2003 Sep 19

[16] Li Y, Park SY, Zhu J. Solid-state anaerobic digestion for methane production from organic waste. Renewable and Sustainable Energy Reviews. 2011 Jan 31;**15**(1):821-826

[17] Khalid A, Arshad M, Anjum M, Mahmood T, Dawson L. The anaerobic digestion of solid organic waste. Waste Management. 2011 Aug 31;**31**(8):1737-1744

[18] Arsova L. Anaerobic digestion of food waste: Current status, problems and an alternative product [Doctoral dissertation]. New York, NY: Columbia University. May 2010

[19] Mata-Alvarez J, editor. Biomethanization of the Organic Fraction of Municipal Solid Wastes. Londond, UK: IWA publishing; 2002 Aug 31

[20] Ostrem K, Themelis NJ. Greening waste: Anaerobic digestion for treating the organic fraction of municipal solid wastes [Master thesis]. New York, NY: Earth Engineering Center Columbia University. 2004. Available from: http://www.seas.columbia.edu/earth/wtert/sofos/Ostrem_Thesis_final.pdf [Accessed: December 5, 2016]

[21] Fernández-Rodríguez J, Pérez M, Romero LI. Comparison of mesophilic and thermophilic dry anaerobic digestion of OFMSW: Kinetic analysis. Chemical Engineering Journal. 2013 Oct 31;**232**:59-64

[22] Kim JK, Oh BR, Chun YN, Kim SW. Effects of temperature and hydraulic retention time on anaerobic digestion of food waste. Journal of Bioscience and Bioengineering. 2006 Oct 31;**102**(4):328-332

[23] Nopharatana A, Pullammanappallil PC, Clarke WP. Kinetics and dynamic modelling of batch anaerobic digestion of municipal solid waste in a stirred reactor. Waste Management. 2007 Dec 31;**27**(5):595-603

[24] Alvarez-Gallego C. Testing different procedures for the startup of a dry anaerobic Co-digestion process of OFMSW and sewage sludge at thermophilic range. [Doctoral Thesis]. Cádiz, Spain: University of Cádiz; 2005

[25] Fdez-Güelfo LA, Álvarez-Gallego C, Márquez DS, García LR. The effect of different pretreatments on biomethanation kinetics of industrial Organic Fraction of Municipal Solid Wastes (OFMSW). Chemical Engineering Journal. 2011 Jul 1;**171**(2):411-417

[26] Fernández J, Pérez M, Romero LI. Kinetics of mesophilic anaerobic digestion of the organic fraction of municipal solid waste: Influence of initial total solid concentration. Bioresource Technology. 2010 Aug 31;**101**(16):6322-6328

[27] Fdez-Güelfo LA, Álvarez-Gallego C, Sales D, García LR. Dry-thermophilic anaerobic digestion of organic fraction of municipal solid waste: Methane production modeling. Waste Management. 2012 Mar 31;**32**(3):382-388

[28] Fernández J, Pérez M, Romero LI. Effect of substrate concentration on dry mesophilic anaerobic digestion of organic fraction of municipal solid waste (OFMSW). Bioresource Technology. 2008 Sep 30;**99**(14):6075-6080

[29] Hartmann H, Ahring BK. Anaerobic digestion of the organic fraction of municipal solid waste: Influence of co-digestion with manure. Water Research. 2005 Apr 30;**39**(8):1543-1552

[30] Lay JJ, Li YY, Noike T. Influences of pH and moisture content on the methane production in high-solids sludge digestion. Water Research. 1997 Jun 1;**31**(6):1518-1524

[31] Hernández-Berriel MC, Márquez-Benavides L, González-Pérez DJ, Buenrostro-Delgado O. The effect of moisture regimes on the anaerobic degradation of municipal solid waste from Metepec (Mexico). Waste Management. 2008 Dec 31;**28**:S14-20

[32] Verma S. Anaerobic digestion of biodegradable organics in municipal solid wastes [Doctoral dissertation]. New York, NY: Columbia University; 2002

[33] Cavinato C, Bolzonella D, Fatone F, Cecchi F, Pavan P. Optimization of two-phase thermophilic anaerobic digestion of biowaste for hydrogen and methane production through reject water recirculation. Bioresource Technology. 2011 Sep 30;**102**(18):8605-8611

[34] Mata-Alvarez J. A dynamic simulation of a two-phase anaerobic digestion system for solid wastes. Biotechnology and Bioengineering. 1987;**30**(7):844-851

[35] Zhang B, Zhang LL, Zhang SC, Shi HZ, Cai WM. The influence of pH on hydrolysis and acidogenesis of kitchen wastes in two-phase anaerobic digestion. Environmental Technology. 2005 Mar 1;**26**(3):329-340

[36] Sreekrishnan TR, Kohli S, Rana V. Enhancement of biogas production from solid substrates using different techniques – A review. Bioresource Technology. 2004 Oct 31;**95**(1):1-10

[37] Bouallagui H, Torrijos M, Godon JJ, Moletta R, Cheikh RB, Touhami Y, Delgenes JP, Hamdi M. Two-phases anaerobic digestion of fruit and vegetable wastes: Bioreactors performance. Biochemical Engineering Journal. 2004 Oct 31;**21**(2):193-197

[38] Igoni AH, Ayotamuno MJ, Eze CL, Ogaji SO, Probert SD. Designs of anaerobic digesters for producing biogas from municipal solid-waste. Applied Energy. 2008 Jun 30;**85**(6):430-438

[39] Sosnowski P, Wieczorek A, Ledakowicz S. Anaerobic co-digestion of sewage sludge and organic fraction of municipal solid wastes. Advances in Environmental Research. 2003 May 31;**7**(3):609-616

[40] Heo NH, Park SC, Kang H. Effects of mixture ratio and hydraulic retention time on single-stage anaerobic co-digestion of food waste and waste activated sludge. Journal of Environmental Science and Health, Part A. 2004 Dec 27;**39**(7):1739-1756

[41] Zhang C, Xiao G, Peng L, Su H, Tan T. The anaerobic co-digestion of food waste and cattle manure. Bioresource Technology. 2013 Feb 28;**129**:170-176

[42] Macias-Corral M, Samani Z, Hanson A, Smith G, Funk P, Yu H, Longworth J. Anaerobic digestion of municipal solid waste and agricultural waste and the effect of co-digestion with dairy cow manure. Bioresource Technology. 2008 Nov 30;**99**(17):8288-8293

[43] Zhang P, Zeng G, Zhang G, Li Y, Zhang B, Fan M. Anaerobic co-digestion of biosolids and organic fraction of municipal solid waste by sequencing batch process. Fuel Processing Technology. 2008 Apr 30;**89**(4):485-489

[44] Dai X, Duan N, Dong B, Dai L. High-solids anaerobic co-digestion of sewage sludge and food waste in comparison with mono digestions: Stability and performance. Waste Management. 2013 Feb 28;**33**(2):308-316

[45] Ponsá S, Gea T, Sánchez A. Anaerobic co-digestion of the organic fraction of municipal solid waste with several pure organic co-substrates. Biosystems Engineering. 2011 Apr 30;**108**(4):352-360

[46] Fricke K, Santen H, Wallmann R, Hüttner A, Dichtl N. Operating problems in anaerobic digestion plants resulting from nitrogen in MSW. Waste Management. 2007 Dec 31;**27**(1):30-43

[47] Rajagopal R, Massé DI, Singh G. A critical review on inhibition of anaerobic digestion process by excess ammonia. Bioresource Technology. 2013 Sep 30;**143**:632-641

[48] Dhar H, Kumar P, Kumar S, Mukherjee S, Vaidya AN. Effect of organic loading rate during anaerobic digestion of municipal solid waste. Bioresource Technology. 2016 Oct 31;**217**:56-61

[49] Gallert C, Winter J. Mesophilic and thermophilic anaerobic digestion of source-sorted organic wastes: Effect of ammonia on glucose degradation and methane production. Applied Microbiology and Biotechnology. 1997 Sep 16;**48**(3):405-410

[50] Ariunbaatar J, Panico A, Esposito G, Pirozzi F, Lens PN. Pretreatment methods to enhance anaerobic digestion of organic solid waste. Applied Energy. 2014 Jun 15;**123**:143-156

[51] Izumi K, Okishio YK, Nagao N, Niwa C, Yamamoto S, Toda T. Effects of particle size on anaerobic digestion of food waste. International Biodeterioration & Biodegradation. 2010 Oct 31;**64**(7):601-608

[52] Zhu B, Gikas P, Zhang R, Lord J, Jenkins B, Li X. Characteristics and biogas production potential of municipal solid wastes pretreated with a rotary drum reactor. Bioresource Technology. 2009 Feb 28;**100**(3):1122-1129

[53] Hansen TL, la Cour Jansen J, Davidsson Å, Christensen TH. Effects of pre-treatment technologies on quantity and quality of source-sorted municipal organic waste for biogas recovery. Waste Management. 2007 Dec 31;**27**(3):398-405

[54] Torres ML, Lloréns MD. Effect of alkaline pretreatment on anaerobic digestion of solid wastes. Waste Management. 2008 Nov 30;**28**(11):2229-2234

[55] Cesaro A, Belgiorno V. Sonolysis and ozonation as pretreatment for anaerobic digestion of solid organic waste. Ultrasonics Sonochemistry. 2013 May 31;**20**(3):931-936

[56] Stamatelatou K, Antonopoulou G, Lyberatos G. Production of biogas via anaerobic digestion. Handbook of biofuels production: Processes and Technologies. Cambridge, UK: Woodhead Publishing Ltd. 2011:266-304

[57] Charles W, Walker L, Cord-Ruwisch R. Effect of pre-aeration and inoculum on the start-up of batch thermophilic anaerobic digestion of municipal solid waste. Bioresource Technology. 2009 Apr 30;**100**(8):2329-2335

[58] Liu X, Wang W, Gao X, Zhou Y, Shen R. Effect of thermal pretreatment on the physical and chemical properties of municipal biomass waste. Waste Management. 2012 Feb 29;**32**(2):249-255

[59] Shahriari H, Warith M, Hamoda M, Kennedy KJ. Anaerobic digestion of organic fraction of municipal solid waste combining two pretreatment modalities, high temperature microwave and hydrogen peroxide. Waste Management. 2012 Jan 31;**32**(1):41-52

[60] Ghosh C. Integrated vermi-pisciculture – An alternative option for recycling of solid municipal waste in rural India. Bioresource Technology. 2004 May 31;**93**(1):71-75

[61] Li WB, Yao J, Tao PP, Hu H, Fang CR, Shen DS. An innovative combined on-site process for the remote rural solid waste treatment – A pilot scale case study in China. Bioresource Technology. 2011 Mar 31;**102**(5):4117-4123

[62] Lu W, Wang H. Role of rural solid waste management in non-point source pollution control of Dianchi Lake catchments, China. Frontiers of Environmental Science & Engineering in China. 2008 Mar 1;**2**(1):15-23

[63] Braun R. Anaerobic digestion: A multi-faceted process for energy, environmental management and rural development. In Improvement of Crop Plants for Industrial end Uses. Netherlands: Springer; 2007. pp. 335-416

[64] Resch C, Braun R, Kirchmayr R. The influence of energy crop substrates on the mass-flow analysis and the residual methane potential at a rural anaerobic digestion plant. Water Science and Technology. 2008 Jan 1;**57**(1):73-81

[65] Rao PV, Baral SS, Dey R, Mutnuri S. Biogas generation potential by anaerobic digestion for sustainable energy development in India. Renewable and Sustainable Energy Reviews. 2010 Sep 30;**14**(7):2086-2094

[66] Bruun S, Jensen LS, Sommer S. Small-scale household biogas digesters: An option for global warming mitigation or a potential climate bomb? Renewable and Sustainable Energy Reviews. 2014 May 31;**33**:736-741

[67] Chen Y, Yang G, Sweeney S, Feng Y. Household biogas use in rural China: A study of opportunities and constraints. Renewable and Sustainable Energy Reviews. 2010 Jan 31;**14**(1):545-549

[68] Jingura RM, Matengaifa R. Optimization of biogas production by anaerobic digestion for sustainable energy development in Zimbabwe. Renewable and Sustainable Energy Reviews. 2009 Jun 30;**13**(5):1116-1120

Chapter 8

Palm Oil Mill Solid Waste Generation and Uses in Rural Area in Benin Republic: Retrospection and Future Outlook

Tatiana W. Koura, Gustave D. Dagbenonbakin,
Valentin M. Kindomihou and Brice A. Sinsin

Additional information is available at the end of the chapter

http://dx.doi.org/10.5772/intechopen.70033

Abstract

Palm oil is one of the major oil crops in the world, producing important vegetable oils in the world oil and fats market. Its production generates solid wastes whose sustainable management is crucial for the oil chain development in oil palm producing countries. Benin Republic is a small oil palm producing country where oil palm plays social, cultural, and economic roles for farmers. This chapter analyzes the linkage between improvement of palm oil process extraction and palm oil mill solid waste (POMSW) management for sustainable palm oil production. Composed mainly of fibers, the two kinds of POMSW are empty fruit bunches (EFBs) and press mesocarp fibers (PMFs), which are rich in units' fertilizers and are renewable energy. POMSW in Benin Republic is used in agriculture, in cosmetic, or as energy. The upgrade of traditional mills generates POMSW use as a boiler fuel to reducing wood necessity and increasing farm profit. As this use is not sustainable, research must be made to generate electricity with POMSW and its use for crop fertilization, to ensure environment protection, enhance contribution to food security, restore degraded soils, and increase earnings of producers of rural areas.

Keywords: POMSW, improvement of palm oil process extraction, electricity, fertilization, rural area

1. Introduction

Within these last decades, with the population growth and food security, both developed and developing countries face many environmental challenges as waste management [1, 2]. Nowadays, the sustainable management of waste is a global issue, because of their permanent increase and their harmful effects on the environment. According to Sotamenou and Kamgnia [3], wastes are produced during household, agricultural, industrial, and commercial activities. In Benin Republic, a new national waste strategy adopted in 2008, concerns mainly solid household wastes and market wastes [4]. Despite population bury and burn household wastes, the solid waste disposal rate is very low in cities and villages. The systems of collection, evacuation, and treatment being little operational and garbage are evacuated in the side streets and empty plots. The situation is worse in rural areas. The survey demographic and health, conducted in Benin in 2001, evaluated the garbage evacuation rate at 39% in urban and 3% in rural areas [5]. There is no specific national strategy to manage agricultural and industrial wastes in Benin. For agricultural wastes, farmers had to burnt them or return them to the field. Almost all research studies on waste management concern household wastes in urban areas [4, 6–8]. Industrial wastes are mostly come from food processing. Small-scale food industries are important in the rural areas because they generate employment; reduce rural-urban migration, and associated social problems. They are vital to reducing postharvest food losses and increasing food availability [9]. Food processing has traditionally been the domain of women. They had to produce little quantities and manage all wastes quantities. Nowadays, food processing had been improving by introduction of new technologies and engines. This is the case of palm oil production in Benin Republic. Oil palm is an oleaginous crop. It provides 39% of vegetable oil world production with 7% of oleaginous plantation areas compared with soybean (61%), colza (18%), and sunflower (14%) [10]. Benin Republic is a small oil palm producing country, where oil palm plays social, cultural, and economic roles for farmers. In 1848, palm oil gradually replaced slave trade. Oil palm through its products, the palm oil and "sodabi" (local palm wine), highly contributes to the income and social capital accumulation; this also discriminates operators and their households socially and economically. In Southern Benin, the more the oil palm acreage is wide, the more farmers are wealthy [11]. Oil palm is cultivated by many farmers and retailed to secure a decent retirement. They used this crop to affirm and secure their land. In addition, incomes from palm kernel sales help households to pay their children school fees. The local wine is used in festivities and ceremonies (weeding, mourning, receptions, etc.). This made the oil palm a serious component in populations' culture where it is grown [12]. Moreover, during revitalization of this sector by the government and NGO, oil palm become as a cash crop that means "money symbol" and palm oil become a great interest for people in this production chain, who began to produce palm oil by themselves. These people improved the extraction method by introducing engines [13, 14]. According to the type of machine used for palm oil production in a partial or total process, they are categorized into four palm oil mill processes: traditional palm oil process (no machine use), small mechanized or improved palm oil process (integration of digester engine in the process), motorized or modern palm oil process (integration of digester and press engines in the process), and semi-industrial or mechanized palm oil processing (integration of large cookers, presses, digesters, sterilizers, clarifiers, and other facilities in the process) [15] (**Figure 1**). Despite of that, only 40% of national needs in vegetable oils were covered [16]. These improvements consequently increase palm oil mill wastes to an extent that some mills struggle to recycle all quantities produced. These wastes cause environmental nuisances. According to Ojonoma and Nnennaya [17], the sustainability of the palm oil

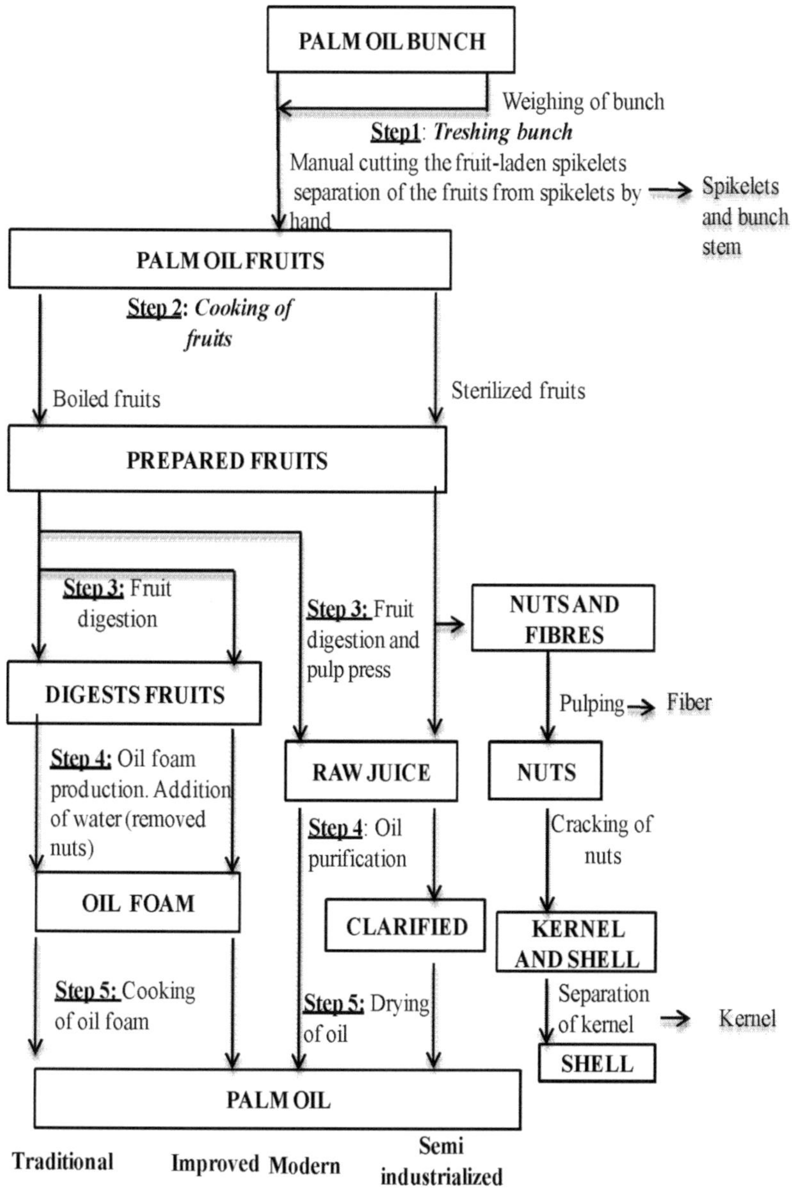

Figure 1. Types of palm oil processing [18].

sector is questioned in the majority of oil palm producing countries because of environmental harm due to the mismanagement of palm oil mill wastes. In Benin Republic, traditional palm oil mills had to use these wastes for many purposes. The present study analyzes the linkage between the improvement of palm oil process extraction and palm oil mill solid waste (POMSW) management for sustainable palm oil production in Benin Republic.

2. Methodology

The data were collected from (i) literature review on the characterization and uses of palm oil mill solid wastes in the world and (ii) research project fieldwork. An approach used for the research project fieldwork was based on the survey in the six departments of the Southern part of the Republic of Benin (**Figure 2**). This part of Benin Republic extends from the coast at 6°25′ to 7°30′ N latitude. It belongs to the Guinea-Congolese zone and submits to subequatorial with two rainy seasons (March–June and September–mid-November) and two dry seasons (July–September and November–March). The annual rainfall of this area, which varies between 1100 and 1400 mm, makes this part of the country adequate for oil palm production. The average daily temperature ranges from 25 to 29°C. The soils are in 66% (700,000 ha) deep lateritic soils of low fertility, and the rest are more fertile alluvial soils and heavy clay soils (360,000 ha) located in the river valleys of Mono, Couffo, Oueme, and in the Lama depression [19]. The survey was carried out from November 2011 to March 2012 and 335 palm oil mills were randomly surveyed using a semi-structure questionnaire. The collected data concerned the method of palm oil production, management of wastes produced, and management practices of waste quantities use (Did he use it? Did he sell it? Did he dump it? How did he use it?). The percentage usage (P_u) or proportion of interviewees who used palm oil mill wastes, the commercial value (CV) or proportion of mills who sell the wastes and rejection rate (RR) or proportion of mills who discard the wastes were calculated as follows:

$$Pu = \frac{\text{Nusers}}{N} \tag{1}$$

where Nusers is the number of informants who use a waste;

$$Cv = \frac{\text{Nw}}{N} \tag{2}$$

where Nw is the number of informants who sell wastes;

$$RR = \frac{\text{Nr}}{N} \tag{3}$$

where *Nr* is the number of informants who discard wastes; and *N* is the total number of informants.

All these parameters vary between 0 and 1.

Concerning POMSW nutrient composition analyses, samples were collected in one semi-industrial palm oil mill. The palm oil extraction process was followed three times and at each time, sample of 1 kg of each kind of wastes was randomly selected. All the samples were mixed and a sample of 500 g was taken. The analyses were performed with ion chromatography system Dionex ICS 1000. Nitrogen was determined by the Kjeldahl method.

Double principal component analysis (PCA) was performed using SAS software to explain the relation between palm oil mill categories and POMSW uses.

http://dx.doi.org/10.5772/intechopen.70033

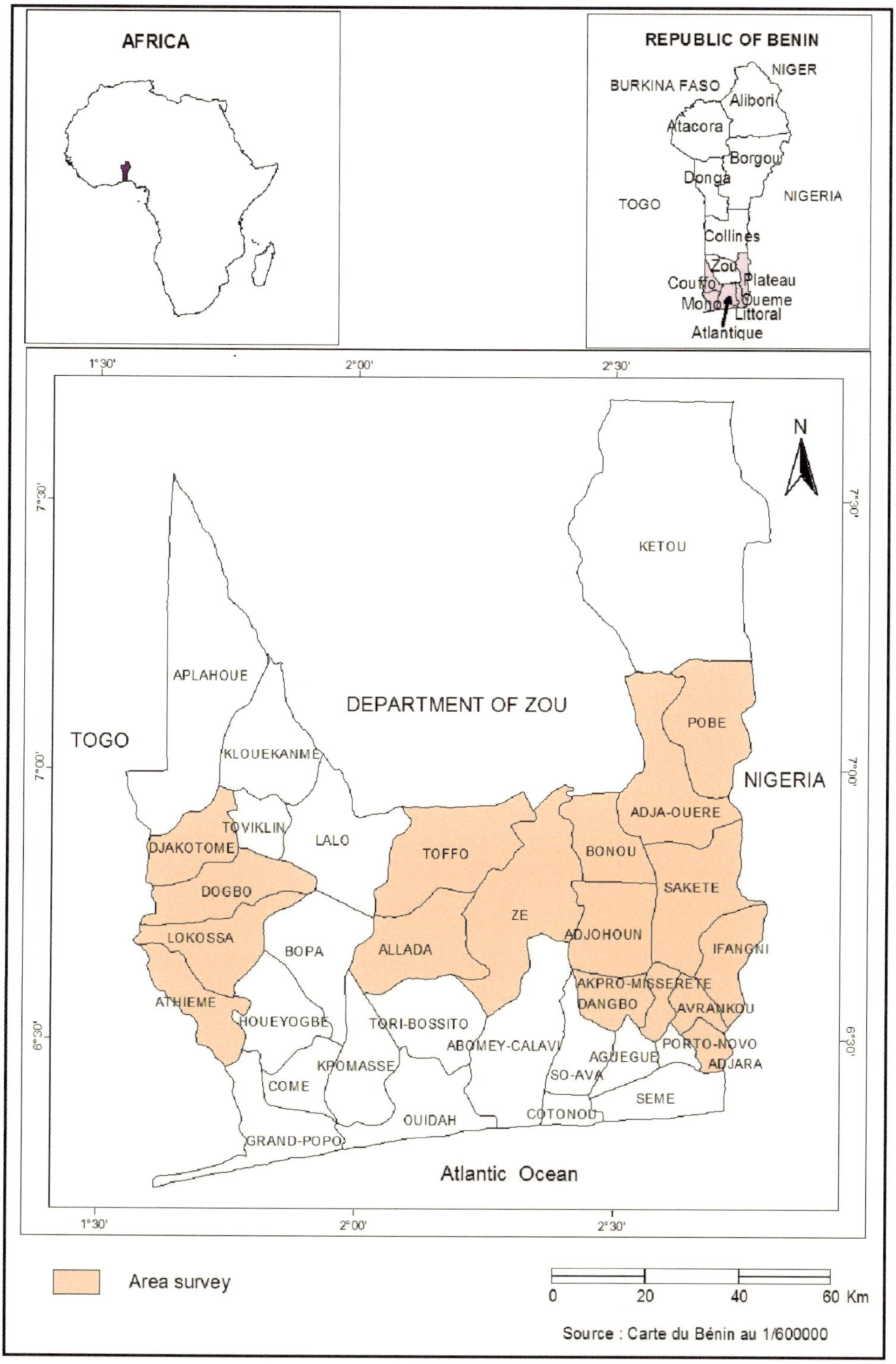

Figure 2. Study area.

3. Results and discussion

3.1. Types, chemical and mineral composition of POMSW

Palm oil mills generate two kind of solid wastes during palm oil fruits transformation: empty fruit bunches (EFBs) and palm mesocarp fiber (PMF) (**Figure 3**). EFB is obtained after the removal of oil seeds from fruit bunches. It is the supporting structures of the oil-bearing fruits in a bunch and comprises spikelet (68–80% dry matter) and stalks (20–32% dry matter) [20, 21]. PMF is the fibrous residues separated from the mesocarp and kernel during palm oil extraction [22]. POMSW is mainly composed of fibers. Spikelet contains more fibers than stalk. Fibers of spikelet are stronger than those of stalk. PMFs are richer in cellulose and lignin than EFB (**Table 1**).

The chemical and mechanical proprieties of these fibers vary with the type of waste. Fibers are the wastes that contain the most ammonium and nitrates. Stalks and spikelet contain low amount of phosphorus. All these wastes contained large amount of calcium (0.6–1.6%) and sulfur (0.2–0.7 mg/g). Fibers and stalks contain large amount of chloride (21.4 and 20.7 mg/g, respectively). The high amount of potassium and chloride can be explained by the fertilizer of oil palm in plantation with KCl.

3.2. Solid biomass from Benin oil palm processing mills in rural area

In Benin Republic, from 1 ton of fresh fruit bunch (FFB), any mill obtains 152.3 l of crude palm oil and generates an average 254.7 kg of EFB and 114.9 kg of PMF [15, 18]. Compare to the other oil palm producer countries (**Table 2**), there are no great differences on EFB obtained. However, mills from Benin produce more EFB than those from Indonesia and less PMF than those from Nigeria,

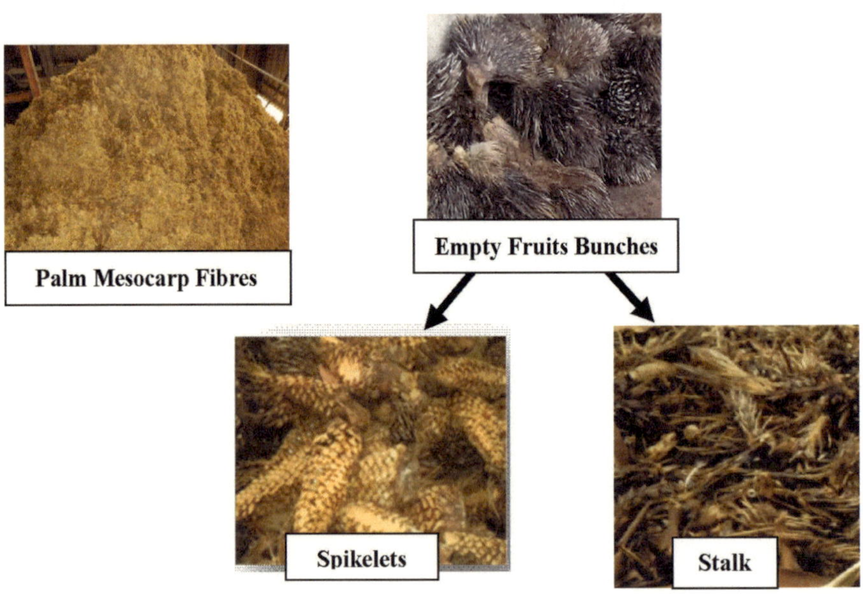

Figure 3. Palm oil mill solid wastes (POMSW). **Source:** Koura pictures.

Main fraction/ Parameters	EFB (literature) [20, 23–30]	Spikelet (study)	Spikelet (literature) [20, 32]	Stalk (study)	Stalk (literature) [31, 32]	PMF (study)	PMF (literature) [33]
C (%)	44.1–54.5	na	50.23–51.67	na	43.62–48.6		45.61
N (%)	0.44–1	0.95	0.5–0.96	1.2	0.7–071	1.4	1.36
C/N	58.9, 77.7	na	–	na	–	50.3	33.54
Lignin (%)	10.5–36.6	na	23.5–29.10	na	–	na	11–20.5
Cellulose (%)	33.7–63	na	20.6–20.7	na	26.9–28.8	na	14–39.9
Hemicellulose (%)	20.1–35.3	na	23.9–28.9	na	24–28.8	na	20.8–28.9
P (%)	0.03–0.7	0.001	0.05–0.19	0.001	0.07–0.3	0.17	–
K (%)	1.4–2.8	16.9	1.75–1.78	16.2	3.31–4.03	4.6	–
Ca (%)	0.16–0.9	0.9	2–2.41	0.6	0.09–0.31	1.5	–
Mg (%)	0.008–0.8	0.003	0.12–0.17	Traces	0.13–0.15	0.9	–
Na (%)	–	0.6	0.001–0.03	0.5	0.004–0.05	1.4	–
Cl^- (mg/g)	–	4.7	–	21.4	–	20.68	–
SO_4^{2-} (mg/g)	0.1–1.4	0.7	–	0.19	–	0.27	–

Table 1. Mineral and chemical composition of POMSW.

Malaysia, and Thailand. These differences can be explained by the variety of fruits used to produce oil. "Dura" variety possesses more shell than the kernel, while "Tenera" possesses more kernel than the shell [34] and the quantities of PMF produce by Dura are less than those from Tenera [35]. In Benin Republic, POMSW quantities vary with the oil extraction process (**Figure 4**).

The semi-industrialized process produced significantly more EFB and PFM than the traditional process. In fact, mills that extracted palm oil by semi-industrialized process used only Tenera fruit variety. Mills that use traditional method transform more Dura variety [18]. The other mills use the two seed varieties. POMSW quantity trends were similar to palm oil produced (**Figure 5**). From 1961 to 1968, 1980 to 1985 and 1991 to 1999, POMSW was relatively stable. After 1968 and 1985, POMSW drop in 1971 and 1997. A pic evolution of POMSW was obtained in 176 (78,600.8 tons of EFB and 35458,3 of PMF. After 1999, POMSW increased quickly from

Country	Benin	Indonesia	Indonesia	Mwalaysia	Nigeria		Thailand	Thailand
References	[15]	[36]	[37]	[38]	[35]		[39]	[40]
Varieties	–	Tenera	Tenera	Tenera	Dura	Tenera	Tenera	Tenera
EFB (kg)	254.7	225	210	230–250	237–324	257–282	240	214–316
PMF (kg)	114.9	143	144	130–150	232–281	191–203	140	120–130
Palm oil (l)	152.3	218	235	160–200	94–128	260–282	–	250–280

Table 2. Palm oil wastes and crude palm oil quantities generated from 1 t of full fruit bunch in different country.

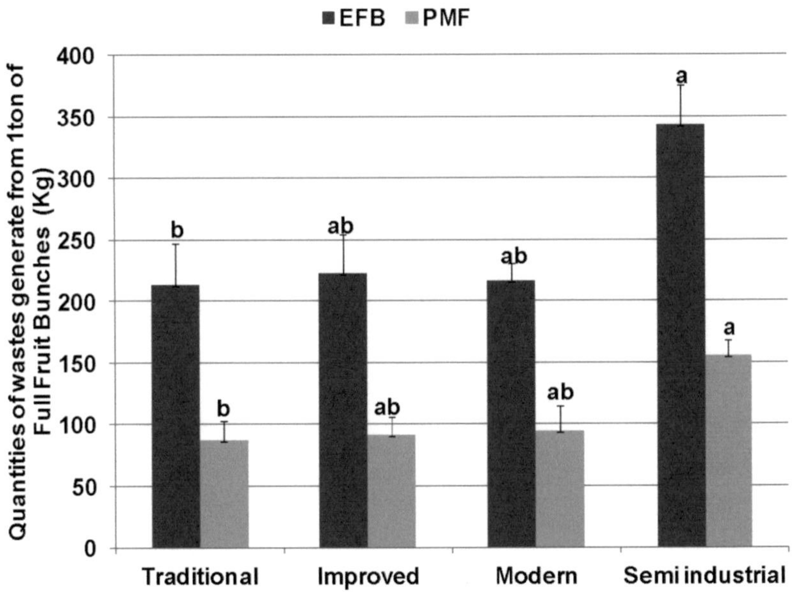

Figure 4. Quantities of POMSW generate by each category of mills. a; b and c: figures followed by different letters are significantly different (Tukey HSD test, p<0.05) (adapted from Ref. [15]).

60,204.86 and 27,159.6 tons in 2000 to 93,652 and 42,248 tons in 2013 for EFB and PMF, respectively. This period corresponds to the entrance of men in palm oil chain value. These men possess large areas of exploitable selected oil palm plantation and have a high financial capacity to buy modern engine or build big palm oil extraction engine and to employ a large number of laborers [13, 14, 18]. In 2022, POMSW quantities generated by mills are projected to reach

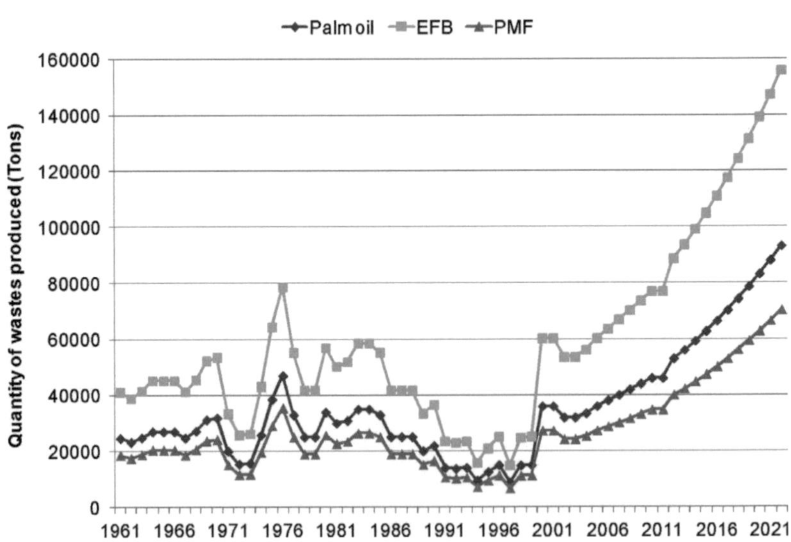

Figure 5. Evolution of POMSW biomass generated in Benin Republic [41].

155,821.3 tons of EFB and 70,294 tons of PMF. Koura et al. [18] identified four classes of oil mills based on the quantity of waste produced: small, medium, large, and very large producers of waste. The analysis of POMSW quantities generated by mills of regional union of oil palm producers (RUOPPs) union régionale des producteurs de palmier à huilie (URPPH) during the last years reveals that EFB and PMF increased only in mills that used the modern and semi-industrialized process (**Figure 6**).

3.3. Palm oil mill solid waste management in sustainability context

In Benin Republic, some mills do not use all of their generated POMSW. Consequently, they sell and/or discard the excess (**Table 3**). The PMFs are more used and sold than EFB. When traditional mill owners decide to upgrade their mills by using the improved extraction method, most of them used these wastes. However, fewer mill owners who practice the modern extraction method use POMSW. Compared to other mill categories, most semi-industrialized mills sell and reject PMF. The problem of fiber and empty fruit bunches management is not related to the amount of waste generated. In fact, palm oil mills are facing problems of management of fiber and empty fruit bunches even if they are produced in small quantities.

3.3.1. Uses of POMSW according to mill categories in the south of Benin Republic

POMSW was used as energy, in agriculture and cosmetic (**Figures 7** and **8**). EFB was burned and the ash was used as potassium in preparation of local soap called "Koto". According to FAO [43], this ash is also used as a fertilizer by some mills. Analyses of the physicochemical parameters of this ash by Udoetok [44] in Nigeria reveal that it contains appreciable amount of plant nutrients such as calcium (146.15 mg/kg), potassium (139.35 mg/kg), nitrate (97.6 mg/kg), phosphate (47.5 mg/kg), sodium (0.63 mg/kg), magnesium (1.68 mg/kg),

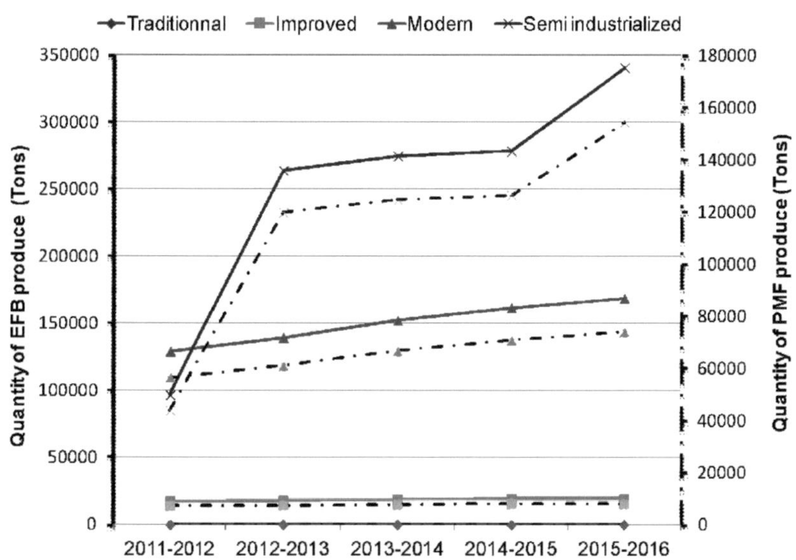

Figure 6. Evolution of POMSW biomass according to extraction palm oil process [42].

	% users (Up)		Commercial value (CV)		Rejection rate (RR)	
Mills categories	**EFB**	**PMF**	**EFB**	**PMF**	**EFB**	**PMF**
Traditional (50)	72	98ab	0.02b	0.1b	0.4	0.1b
Improved (134)	83.6	100a	0b	0.3a	0.5	0.1ab
Modern (142)	87.3	94.4b	0.03b	0.1b	0.4	0.01c
Semi-industrialized (9)	88.9	88.9ab	0.1a	0.6 a	0.4	0.6a
P (0.05)	0.09	0.02	$<2.2e^{-16}$	$3.2\ e^{-6}$	0.37	$3.9e^{-10}$

Notes: Values in bracket are the number of mills surveyed in each category. The *p*-values displayed indicate a significant difference among the mill categories according to each parameter ($p < 0.05$). a, b, and c: figures followed by different letters are significantly different.

Table 3. Management of palm oil mill waste quantity generated according to each mill category.

and zinc (0.38 mg/kg) and that it justifies its use as an organic manure. POMSW was used directly or indirectly in agriculture as the fertilizer. Fresh POMSW was applied in palm plantation (33.1% of informants) by using two methods. The most common method is local application and the second is mulching [15]. Schuchardt et al. [45] stated that EFBs need to be applied in fresh state to reduced erosion, decreased nitrogen losses, controlled weed growth, improved soils nutrients, and avoided the danger of *Ganoderma boninense* and *Oryctes rhinoceros* (rhinoceros beetle) build up, important oil palm pathogen and pest. Bakar et al.'s study [46] shows that the application of 300 kg of POMSW per year in heaps in the middle of every four palms during 10 years improved the soil physicochemical characteristics of the top of soil (0–60 cm) and increased the fresh fruit bunch (FFB) yield more than 150 kg EFB and NPK. Nwoko and Ogunyemi [47], Embrandiri et al. [48], and Kolade et al. [49] stated that these wastes are very rich in nutrients and improve soil fertility and crop growth and yield. However, composting of POMSW is considered as the sustainable method of POMSW use [49–55]. In Benin Republic, composting was less practiced (13.6% of informants) using the heaping method (87.5% of informants) or pig breeds on POMSW (8.3% of informants) or holing POMSW (4.2% of informants) [18]. The compost made from palm oil mill wastes obtained by producers is used in plantations or vegetable production. Koura et al. recommended the use of POMSW and cattle manure compost applied at 10 t/ha for best tomato yield and the use of POMSW and poultry manure composted altogether in covered system and applied at least at 20 t/ha for best amaranth growth and yield production [56, 57]. Sabrina et al. [58] reveal in their study that fresh, composted, and field composted EFB produced phenol compounds, whereas no phenolic compounds were detected in vermicomposted EFB. EFB was also used for mushroom production. POMSW was used in energy production directly as a boiler fuel and PMF was used indirectly as energy after mixed it with palm oil mill effluent for fire starting cake production.

However, POMSW use as a boiler fuel was prohibited in some countries [59] to preserve human and ecological health [60–62]. The double PCA shows that two axes explain 82% of different POMSW uses according to mill categories. **Table 4** shows the coefficients of correlation between the POMSW uses and mill categories and the first two PCA axes.

http://dx.doi.org/10.5772/intechopen.70033

Figure 7. Uses of POMSW as energy and in cosmetic in Benin Republic. **Source:** Koura pictures. (a) Woman in mixing PMF and POME to make fire starting cake; (b) fire starting cake produce in industrial palm oil mill; (c) fire starting cake produce in traditional palm oil mill; (d) use of EFB as boiler fuel; (e) use of PMF as boiler fuel; (f) local soap making with POMSW called "Koto".

This table shows that axis 1 explains modern (modern) and semi-mechanized (minimech) mills and uses of PMF for fertilization (ffert) and fire starting cake production (ffire). The axis 2 explains traditional (traditio) and improved (improved) mills and EFB uses as a boiler fuel (eboil) or mushroom production (emush) and PMF uses as a boiler fuel (fboil). **Figure 9** shows the projection of the POMSW uses by palm oil mill category in the system axes 1 and 2.

Figure 8. Uses of POMSW in agriculture in Benin Republic. **Source:** Koura pictures (2012). (a) Local application of EFB in plantation; (b) mulching system of EFB and PMF application in plantation; (c) heaping POMSW near mills for decomposition; (d) pig breeding on POMSW; (e) old compost obtain from pig breeding system.

The results show that modern mills use more PMF for fire starting production and EFB as a boiler fuel. Mini industrial mills use fibers for fertilization and boiler fuel and EFB for soap production. Traditional mills use EFB for fertilization, mushroom production, and PMF for

Parameters	Axis 1	Axis 2
Traditional	−0.021	0.159
Improved	0.04	−0.06
Modern	−0.059	−0.005
Semi-industries	0.509	−0.102
EFB uses for soap	0.017	−0.015
EFB uses as boiler fuel	0.008	−0.046
EFB uses for fertilization	−0.073	0.174
EFB uses for mushroom production	0.034	0.297
Fiber uses for fertilization	1.084	0.367
Fiber uses as boiler fuel	0.021	−0.034
Fiber uses for fire starting cake production	−0.134	0.06

Table 4. Correlation between the POMSW uses and mill categories and the first two PCA axes (in brackets is the proportion of variation explained by each axis, expressed in percentage).

fertilization and fire starting production. Improved mills use more POMSW as the boiler fuel and sometime EFB for soap production. These wastes can be used for other purposes. According to Abdullah and Sulaiman [63], EFB and PMF are clean, noncarcinogenic, free from pesticides, and soft parenchyma cells. Consequently, they can be used in erosion control, mattress cushion production, soil stabilization, horticulture and landscaping, ceramic and brick manufacturing, paper production, and acoustics control [63]. In great oil palm countries, such as Malaysia, Indonesia, and Thailand, others potentialities of oil palm wastes had been studied [64]. These results demonstrate the possibility of employing hydro thermal for producing solid fuel as well as nutrient recovery from EFB. POMSW can also use as a source of renewable energy [64, 65]. In fact, they can produce steam for processing activities and for generating electricity [64].

3.3.2. Factors that influence palm oil mill waste management in rural area

The wastes management choice must be influencing by many factors. In Garissa municipality, a study shows that understaffing, lack of education, poor supervision, lack of appropriate facilities, and lack of resident's support are among reasons leading to poor solid waste management [66]. In Benin Republic, the use of POMSW by a mill does not depend on waste quantities [18] but on the knowledge of producers on this waste uses, the importance and economical input of these wastes. In fact, it had been shown by Koura et al. [67] that the use values of these wastes depend on their importance for mill owners and by Koura et al. [15] that the uses of POMSW for new purposes as composting depends on farmers' knowledge on what is compost, composting method and possibilities to compost POMSW. Contrary to traditional mills, all others mills use POMSW as the boiler fuel. These mills reduce the quantities of wood use to cook palm fruits with POMSW.

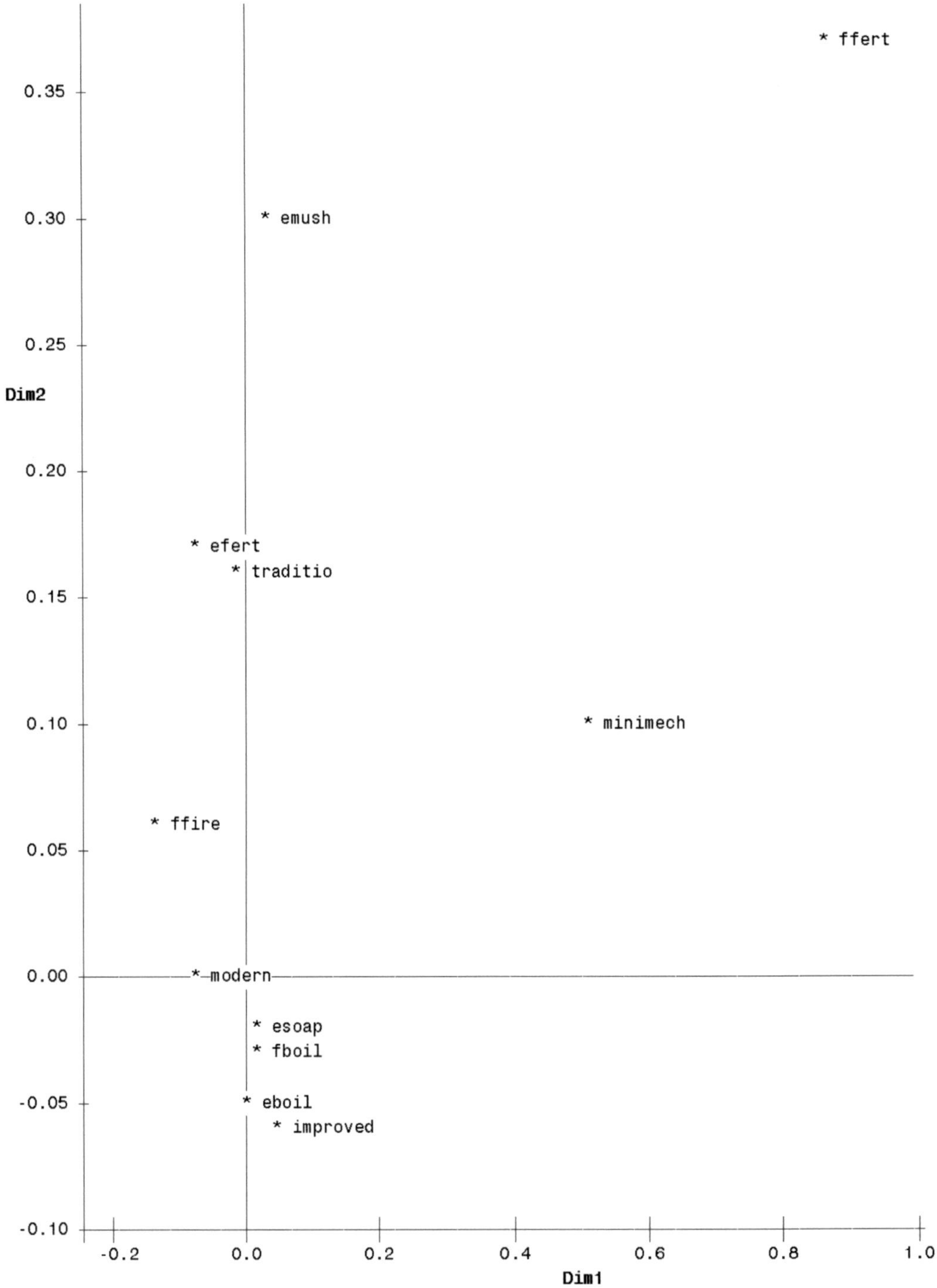

Figure 9. Projection of POMSW uses by mill categories in the system axes 1 and 2.

http://dx.doi.org/10.5772/intechopen.70033

3.4. Future outlook for sustainable POMSW management in rural area

POMSW is useful for farmers in Benin Republic. However, because of the large quantity of wastes produced, some palm oil mills face waste management problem. The choice of one use of these wastes depends on its importance and economic input. By improving oil extraction process, mills are confronted to wood necessity as fuel for stoves and boilers. Consequently, these wastes are priority use as the boiler fuel. This use must be prohibited for environment protection. POMSW is mainly composed of fibers that can be used as renewable energy and solve electricity problems in rural areas. **Table 5** presents the estimation of energy content on POMSW in different oil palm countries. Energy content of these wastes is very lower in Benin Republic than the others palm oil producer's countries. However, this is an opportunity for palm oil mill owners' country, where there is a predominance of wood energy (fuel wood and charcoal) in the national energy balance. Fuel wood represents 59.4% in the final energy total consumption in 2005, while petroleum products accounted for 38.4%. Electricity represents only 2.2% of these intakes [14].

On the other hand, POMSW is agricultural waste, rich in unit fertilizers in particular nitrogen and potassium. The best manner to valorize agricultural wastes is their use as fertilizers to improve soil fertility and increase crop yields, hence enhance food security since waste generation had increased with population expansion and industrialization [72]. The present study reveals that the use of POMSW as fertilizer was practiced in traditional mills and was abandon with the upgrade of traditional mill to improve and modern palm oil mills. This use was practiced in semi-industrial mills that produce big quantities of POMSW. Further research must conduct on the possibilities of using biogas derive from composting to produce electricity for the mill and compost for soils fertilization. It is important to integrate raw materials rich in phosphorus such as poultry manure for POMSW composting because they are poor in this nutrient. These two kinds of POMSW uses will ensure environmental protection, contribute to food security, restore degraded soils, and increase earning money of producers of rural areas.

POMSW	Benin Republic [68]	Nigeria (Rivers state) [69]	Nigeria (Imo state) [70]	Malaysia [71]
EFB (MJ/kg)	4.4	17.75	–	18.84
PMF (MJ/kg)	9.6	18.75	19.67	19.07

Table 5. Energy content of POMSW.

4. Conclusion

Mainly formed of fibers, EFBs composed of spikelet and stalk, and PMFs are the solid wastes generated during palm oil fruit transformation. These wastes contained large amount of nitrogen, calcium, potassium, sulfur, and chloride and less phosphorus. In Benin Republic, POMSW had increased only in mills that used modern and semi-industrialized process during the 5 last years. Some mills sold and/or discarded these wastes.

The present study reveals that as mini industrial mill that produces big POMSW quantities, traditional mills are confronted to waste management. POMSW was used as energy, in agriculture and cosmetic. The upgrade of traditional mill to improve or modern mills creates the need of wood to feed boiler and stoves. However, this use must be avoided preserving environment. Since the use of POMSW depends on its importance and economic input, furthers studies must be made on its use for electricity generation and cropping soil fertilization through composting.

Author details

Tatiana W. Koura[1*], Gustave D. Dagbenonbakin[2], Valentin M. Kindomihou[1,3] and Brice A. Sinsin[1]

*Address all correspondence to: thalia052002@gmail.com

1 Laboratory of Applied Ecology, Faculty of Agronomic Sciences, University of Abomey Calavi, Benin Republic, West Africa

2 Communication and Documentation in Agric Center of Cotton and Fiber Researches, National Institute for Agricultural Research, Benin Republic, West Africa

3 Department of Animal Production, Faculty of Agronomic Sciences, University of Abomey Calavi, Benin Republic, West Africa

References

[1] Yaker F, Sidoum H, Rousselle V, Smaoun S. Le compostage. enda tiers monde – Preceup; 1996. p. 15

[2] FAO. Fruits et légumes biologiques des régions tropicales. In Conférence des Nations Unies pour le Commerce et le Développement; 2003. pp. 170-175

[3] Sotamenou J, Kamgnia DB. La décentralisation pour une gestion efficace des déchets solides municipaux de la ville de Yaoundé. Faculté des sciences économiques et de gestion Université de Yaoundé II Soa; 2009

[4] Ngahane EL. Gestion technique de l'environnement d'une ville (Bembereke au BENIN): caractérisation et quantification des déchets solides émis; connaissance des ressources en eau et approche technique [Thesis]. Département des Sciences et Gestion de l'Environnement. Faculté des Sciences. Université de Liège; 2015

[5] Fonds Africain de développement. Rapport d'évaluation. Programma d'alimentation en eau potable et d'assainissement en milieu rural en république du Bénin; 2014. p. 77

[6] Hounkpatin R, Kottin MC. La gestion des déchets solides ménagers (DSM) à Cotonou: proposition d'un cadre approprié de planification de la pré-collecte. [Memory] ENEAM. University of Abomey-Calavi; 2009

[7] Roch Gbinlo. Organisation et lancement de la gestion des déchets ménagers dans les villes de l'Afrique Subsaharienne: le cas de la ville de Cotonou au Bénin. Economies et finances. Université d'Orléans; 2010

[8] Tokpanou A. Gestion des déchets solides ménagers dans la ville d'Abomey-Calavi (Bénin): Caractérisation et essais de valorisation par compostage [thesis]. Chimie de l'environnement-Chimie des déchets. Université d'Abomey-Calavi/Université d'Aix marseille; 2012

[9] Aworh OC. The role of traditional food processing technologies in National Development: The West African Experience. In: Robertson GL, Lupien JR, editors. Using Food Science and Technology to Improve Nutrition and Promote National Development. International Union of Food Science and Technology; 2008. pp. 1-18

[10] Rival A. The oil palm oil: Challenges and questions to research. Oléagineux, Corps Gras, Lipides. 2013;**20**(3):133-142. DOI: 10.1051/ocl.2013.0506

[11] Adegbola YP, Sodjinou E, Akoha S. Diagnostic des contraintes à la production cotonnière au Bénin. Cotonou: Institut National de Recherche Agricole-Bénin (INRAB); 2009. p. 14

[12] Fournier S, Muchnik J, Requier-Desjardins D. Enjeux et contraintes du développement de la filière huile de palme au Bénin: une approche par les systèmes agro-alimentaires localisés. Les Cahiers d'Outre-Mer; 2002;**220**:475-494

[13] Fournier S, Ay P, Jannot C, Okounlola-Biaou A, Pédé E. La transformation artisanale de l'huile de palme au Benin et au Nigeria. Montpellier: Cirad; 2001. p. 134

[14] Hodonou A. Les filières agricoles: écueils, leçons et perspectives: cas du palmier à huile et du coton. Projet de renforcement des capacités en conception et analyse des politiques de développement; 2010. p. 20

[15] Koura WT, Dagbenonbakin DG, Kindomihou MV, Sinsin AB. Farmers' background and diversity of uses of Palm oil Wastes for sustainable agriculture in Southern Benin Republic. Biological Agriculture and Horticulture Journal. 2014;**31**(1):35-44. DOI: 10.1080/01448765.2014.964316

[16] MAEP (Ministère de l'Agriculture, de l'Élevage et de la Pêche). Strategic plan to revival agricultural sector: Cotonou (Benin); MAEP; 2011. p.116

[17] Ojonoma OL, Nnennaya R. The environmental impact of palm oil mill effluent (POME) on some physic-chemical parameters and total aerobic bioload of soil at a dump site in Anyigba, Kogi state, Nigeria. African Journal of Agriculture Research. 2007;**2**(12):656-662

[18] Koura WT, Kindomihou MV, Dagbenonbakin DG, Janssen M, Sinsin AB. Quantitative assessment of palm oil mill wastes generated by mills in Southern Benin. African Journal of Agricultural Research. 2016;**11**(19):1787-1796. DOI: 10.5897/AJAR2013.8124

[19] Adjanohoun EJ, Adjakidje V, Ahyi MRA, Ake Assi L, Akoegninou A, d''Almeida J, Apovo F, Bouke FK, Chadare M, Cusset G, Dramane K, Eyme J, Gassjta JN, Gbaguidi

N, Goudote E, Guinko S, Houngnon P, Issa LO, Keita A, Kiniffo HV, Konebamba D, Musampa Nseyya A, Saadou M, Sodogandji T, De Souza S, Tchabi A, Zinsou Dossa C, Zohoun T. Contribution aux études ethnobotaniques et floristiques en République Populaire du Bénin. Médecine traditionnelle et pharmacopée. ACCT; 1989. p. 895

[20] Harun MH, NoorMd MR. Fruit set and oil palm bunch components. Journal of Oil Palm Research. 2012;**14**(2):24-33

[21] Nafu YR, Foba-Tendo J, Njeugna E, Oliver G, Cooke KO. Extraction and characterization of fibres from the stalk and spikelets of empty fruit bunch. Journal of Applied Chemistry. 2015;**2015**:10. DOI: http://dx.doi.org/10.1155/2015/750818

[22] Nordin NIAA, Ariffin H, Andou Y, Hassan MA, Shirai Y, Nishida H, Yunus WMZW, Karuppuchamy S, Ibrahim NA. Modification of oil palm mesocarp fiber characteristics using superheated steam treatment. Molecules. 2013;**18**(8):9132-9146. DOI: 10.3390/molecules18089132

[23] Razali Wan WA, Baharuddin AS, Talib AT, Sulaiman A, Naim MN, Hassan MA, Shirai Y. Degradation of oil palm empty fruit bunches (OPEFB) fibre during composting process using in Vessel composter. Bioresources. 2012;**7**(4):4786-4805

[24] Zainudin NF, Lee KT, Kamaruddin AH, Bhatia S, Mohamed AR. Study of adsorbent prepared from oil palm ash (OPA) for flue gas desulphurization. Separation and Purification Technology. 2005;**45**:50-60

[25] Law KN, Wan Daud WR, Ghazali A. Morphological and chemical nature of fiber tenders strands of oil palm empty fruit bunch (OPEFB). Bio Ressources. 2007;**2**(3):351-362

[26] Kelly-young TL, Lee KT, Mohamed AR, Bhatia S. Potential of hydrogen from oil palm biomass as a source of renewable energy worldwide. Energy Policy. 2007;**35**:5692-5701. DOI: 10.1016/j.empl.2007.06.017

[27] Baharuddin AS, Sulaiman A, Kim DH, Mkhtar MN, Hassan MA, Wakisaka M, Shirai Y, Nishida H. Selective component degradation of oil palm empty fruit bunches (OPEFB) using high-pressure steam. Biomass & Bioenergy. 2013;**55**:268-275. DOI: 10.1016/j.biombioe.2013.02.013

[28] Mohammad N, Alam MZ, Kabbashi NA, Ahsan A. Effecting composting of oil palm industrial waste by filamentous fungi: A review. Resource, Conservation & Recycling. 2012;**58**:69-78

[29] Baharuddin AS, Wakisaka M, Shirai Y, Abd-Aziz S, Abdul Rahman NA, Hassan MA. Co-composting of empty fruit bunches and partially treated palm oil mill effluents in pilot scale. International Journal of Agricultural Research. 2009;**4**(2):69-78. DOI: 10.3923/ijar.2009.69.78

[30] Kavitha B, Jothimani P, Rajannan G. Empty fruit bunch: A potential organic manure for agriculture. International Journal of Science, Environment and Technology. 2013;**2**(5):930-937

[31] Zaharah AR, Lim KC. Oil empty fruit bunch as a source of nutrients and soil ameliorant in oil palm plantations. Malaysian Journal of Soil Science. 2000;**4**:51-66

[32] Yunos Md NSH, Baharuddin AS, Yunos Md KF, Hafid HS, Busu Z, Mokhtar MN, Sulaiman A, Som Md A. The physicochemical characteristics of residual oil and fibers from oil palm empty fruit bunches. Bioresources. 2015;**10**(1):14-29

[33] Koba Y, Ayaaki I. Chemical composition of palm fiber and its feasibility as cellulosic raw material for sugar production. Agricultural and Biological Chemistry. 1990;**54**(5):1183-1187. ISSN: 0002-1369

[34] Jimoh MO, Olukunle OJ. Effect of physico-mechanical properties of palm nut on machine performance evaluation. World Applied Programming. 2013;**3**(7):302-308

[35] Ohimain EI, Izah SC, Obieze FAU. Material-mass balance of smallholder oil palm processing in the Niger Delta, Nigeria. Advance Journal Food Sciences Technology. 2013;**5**:289-294

[36] Hayashi K. Environmental impact of palm oil industries in Indonesia. In: Proceedings of International Symposium on EcoTopia Science; 23-25 November 2007; Nagoya, Japan. Nagoya University; 2007. pp. 646-651

[37] Hambali E, Thahar A, Komarudin A. The potential oil palm and rice biomass as bioenergy feedstock. In: 7th Biomass Asia Workshop; 29th November 2009-01st December 2010; Jakarta, Indonesia; 2010

[38] Maheswaran A, Singam G. Pollution control in the palm oil industry—Promulgation of regulations. Planter. 1977;**53**:470-476

[39] Chavalparit O, Rulkens WH, Mol APJ, Khaodhair S. Options for environmental sustainability of the crude palm oil industry in Thailand through enhancement of industrial ecosystem. Environmental, Development and Sustainability. 2006;**8**:271-287

[40] Prasertsan S, Prasertsan P. Biomass residues from palm oil mill in Thailand: An overview on quantity and potential usage. Biomass & Bioenergy. 1996;**11**(5):387-395

[41] Actualitix. Huile de palme—Pays producteurs (Tonnes). Available from: http://fr.actualitix.com/pays/wld/huile-de-palme-pays-producteurs.php [Accessed: February 02, 2017]

[42] Koura WT. Etat des lieux de production d'huile de palme au Bénin. Rapport d'enquête; 2017. p. 60

[43] FAO. Small scale palm oil processing in Africa. FAO Agricultural Services Bulletins N°148; 2002. p. 66

[44] Udoetok IA. Characterization of ash made from oil palm empty fruit bunches (OEFB). International Journal of Environmental Sciences. 2012;**3**(1):519-524

[45] Schuchardt F, Wulfert K, Darnoko D, Herawan T. Effect of new palm oil mill processes on the EFB and POME utilization. In: Proceedings of Chemistry and Technology Conference PIPOC 2007; 26-30 August 2007; Kuala Lumpur; p. 14

[46] Bakar RA, Darus SZ, Kulaseharan S, Jamaluddin N. Effects of ten year application of empty fruit bunches in an oil palm plantation on soil chemical properties. Nutrient Cycling in Agroecosystems. 2011;**89**(3):341-349

[47] Nwoko CO, Ogunyemi S. Evaluation of palm oil mill effluent to maize (*Zea mays* L.) crop: Yields, tissue nutrient content and residual soil chemical properties. Australian Journal of Crop Science. 2010;**4**(1):16-22

[48] Embrandiri A, Singh RP, Ibrahim MH, Ramli AA. Land application of biomass residue generated from palm oil processing: Its potential benefits and threats. The Environmentalist. 2012;**32**:111-117

[49] Kolade OO, Coker AO, Sridhar MKC, Adeoye GO. Palm kernel waste management through composting and crop production. Journal of Environment Health Research. 2012;**5**:81-85

[50] Yusri A, Rasol AM, Mohammed O, Azizah H, Kume T, Hashimoto S. Biodegradation of oil palm empty fruit bunch into compost by composite micro-organisms. In: Proceedings of the EU-ASEAN Conference Combustion of Solids and Treated Product; 16-17 February 1995; Hua-Hin; 1995. pp. 8-9

[51] Thambirajah JJ, Zulkifli MD, Hashim MA. Microbiological and biochemical changes during the composting of oil palm empty fruit bunches. Effect of nitrogen supplementation on the substrate. Bioresource Technology. 1995;**52**:133-144

[52] Hock LS, Baharuddin AS, Ahmad MN, Shah UKM, Rahman NAA, Suraini AA, Hassan MA, Shirai Y. Physicochemical changes in windrow co-composting process of oil palm mesocarp fibre and palm oil mill effluent anaerobic sludge. Australian Journal Basic Applied Sciences. 2009;**3**:2809-2816

[53] Rupani PF, Singh RP, Ibrahim MH, Esa N. Review of current palm oil mill effluent (POME) treatment methods: Vermicomposting as a sustainable practice. World Applied Sciences Journal. 2010;**10**(10):1190-1200

[54] Singh RP, Ibrahim M, Esa N, Iliyana M. Composting of waste from palm oil mill: A sustainable waste management practice. Reviews in Environnemental Science and Biotechnology. 2010;**9**(4):331-344

[55] Singh RP, Embrandiri MH, Ibrahim M, Esa N. Management of biomass residues generated from oil mill: Vermicomposting a sustainable option. Resources Conservation and Recycling. 2011;**55**:423-434

[56] Koura WT, Dagbenonbakin DG, Kindomihou MV, Sinsin AB. Effect of composting of palm oil mill wastes and cow dung or poultry manure on *Lycopersicon esculentum* (Mill.) growth and yield. Journal of Organic Agriculture and Environment. 2015;**3**:13-18

[57] Koura WT, Dagbenonbakin DG, Kindomihou MV, Sinsin AB. Effect of co composting of palm oil mill wastes and cow dung or poultry manure on Amaranthus hybridus growth and yield. Journal of Applied Biosciences. 2015;**86**:7918-7927. DOI: http://dx.doi.org/10.4314/jab.v86i1.1

[58] Sabrina DT, Hanafi MM, Muhmud TMM, Rahman AA, Nor AA. Vermicomposting of oil palm empty fruit bunch and its potential in supplying of nutrients for crop growth. Compost Sciences Utilization. 2009;**17**:61-67

[59] Astimar AA, Wahid MB. Supply outlook of oil palm biomass in Malaysia. In: Proceeding of the Seminar on Ecomat Research and Promotion; 24-25 July; Beijing, China; 2006

[60] Hegg DA, Radke LF, Hobbs PV, Brock CA, Riggan PJ. Nitrogen and Sulphur emissions from the burning of forest products near large urban areas. Journal of Geophysical Research. 1987;**92**:14701-14709

[61] Lacaux JP, Loemba-Ndembi J, Lefeivre B, Cros B, Delmas R. Biogenic emissions and biomass burning influences on the chemistry of the fogwater and stratiform precipitations in the African equatorial forest. Sciences de la vie, de la terre et agronomie. Atmospheric Environment. 1992;**26**:541-551

[62] Ezcurra AI, Ortiz de Zarate P, Vhan D, Lacaux JP. Cereal waste burning pollution observed in the town of Vitoria (northern Spain). Atmospheric Environment. 2001;**35**:1377-1386

[63] Abdullah N, Sulaiman F. The oil palm Wastes in Malaysia. Biomass Now—Sustainable Growth and Use. 2013. pp. 75-100. DOI: http://dx.doi.org/10.5772/55302

[64] Nurdiawati A, Noviantia S, Zainia IN, Nakhshinievaa B, Hiroaki S, Takahashia F, Kunio Y. Evaluation of hydrothermal treatment of empty fruit bunch for solid fuel and liquid organic fertilizer co-production. Energy Procedia. 2015;**79**:226-232

[65] Izah SC, Ohimain EI, Angaye TCN. Potential thermal energy from palm oil processing solid wastes in Nigeria: Mills consumption and surplus quantification. British Journal of Renewable Energy. 2016;**01**(01):39-45

[66] Ibrahim SO. Factors Influencing Solid Waste Management in Garissa Municipality Kenya. University of Nairobi; 2012

[67] Koura WT, Dagbenonbakin DG, Kindomihou MV, Phill H, Sinsin AB. Palm oil mill waste importance and its management in a sustainability context in Southern Benin. Revue du CAMES, option Sciences de la vie, de la terre et Agronomie. 2014;**2**(1):50-57

[68] Adrien BIO Yatokpa AB, Mahman S, Abble K. Développer le Bénin à partir des sources d'énergies renouvelables: Identification et cartographie des potentialités et sources d'énergies renouvelables assorties des possibilités d'exploitation. Ministère de l'Energie et de l'eau/PNUD; 2010

[69] Ohimain EI, Izah SC. Energy self-sufficiency of smallholder oil palm processing in Nigeria. Renewable Energy. 2014;**63**:426-431

[70] Sumathi S, Chai SP, Mohamed AR. Utilization of oil palm as a source of renewable energy in Malaysia. Renewable and Sustainable Energy Review. 2008;**12**:2404-2421

[71] Ubabuike UH. Energy and fuel value analysis of ADA palm oil plantation limited, Imo state, Nigeria. American Journal of Mechanical Engineering. 2013;**1**(4):89-95

[72] Sabiiti E. Utilizing agricultural waste to enhance food security. African Journal of Food, Agriculture, Nutrition and Development. 2011;**11**(6):1-9

The Solid Wastes of Coffee Production and of Olive Oil Extraction: Management Perspectives in Rural Areas

Maria Cristina Echeverria, Elisa Pellegrino and Marco Nuti

Additional information is available at the end of the chapter

http://dx.doi.org/10.5772/intechopen.69427

Abstract

There are two problematic solid residues from agriculture and agro-industry, produced in vast amounts in rural areas: those from coffee bean production and processing and those deriving from the extraction process of olive oil. Notwithstanding these residues originating in different geographical areas, they have striking similarities. They both derive from traditional, conventional and organic agriculture; they have a high content in lignins, celluloses and (poly)phenols; they are produced in million tonnes annually; they pose relevant environmental problems for disposal; they contain bioactive compounds; and the approach for their re-use is often similar, sometimes overlapping. The most promising re-uses in rural areas are for agriculture, as animal feed and for energy production. There are also minor uses, suitable for the production of added-value commodities. The re-use will be dependent on a variety of factors according to the diversity of (a) pedoclimatic areas that include altitude and latitude, soil texture and organic matter content, water regime and availability, (b) level of expertise of the small farmers, (c) social environment that includes training opportunities and availability to create associative forms among producers, (d) access to trade and communication networks and (e) easy access to community-level processing installations. The perspectives of agronomic management and valorization are compatible with the objectives of a regenerative, sustainable agriculture.

Keywords: rural areas, coffee solid waste, olive oil extraction solid waste, re-use of agricultural waste, rural areas, regenerative agriculture, valorization of solid residues

1. Introduction

The rural areas where coffee is cultivated are located mainly in the equatorial and sub-equatorial zone in Africa, Asia and South America. The top ten producing countries are Guatemala

(224,871 tonnes/year), Mexico (257,940), Uganda (314,489), Honduras (380,296), India (385,786), Ethiopia (423,287), Indonesia (814,629), Colombia (892,871), Vietnam (1,818,811) and Brazil (2,859,502) [1]. Coffee production and processing, the latter step often taking place in locations distant from the beans production sites, generate yearly over 20 million tonnes of liquid and solid waste to be disposed of by farmers and processing plants. Although the farms in these areas range in size from 0.5 to 6 hectares (ha), typically they are 1–2 ha and coffee is generally grown along with other cash and food crops, such as maize, as well as cattle. An example is provided by coffee farmers in Uganda [2] having larger-than-average farm plots, while farmers growing coffee and maize tend to have larger plots than coffee farmers without maize (2.69 ha compared to 1.86 ha, respectively). Thus, coffee and maize production is a key determinant for household incomes and poverty, and some land is dedicated to traditional staple food with low-added value. Indeed coffee and maize producers have significantly lower poverty rates compared to coffee farmers that do not grow maize. In general, we can assume that coffee farm economics is dependent upon a wide variety of factors, including productivity, quality, costs of production and waste disposal, price premiums, the latter to achieve quality or sustainability standards. The options for the valorization of the coffee residues, not focusing on waste management in rural areas, have been recently reviewed [1, 3–6].

The rural areas where olive trees are cultivated are located mainly in the Mediterranean basin, where in the northern side over 82.5% of the world production of olive oil (i.e. 2.34 out of 2.84 million tonnes) takes places (Spain, Italy, Greece, Portugal, France, Cyprus, Slovenia and Malta) [7]. This figure rises up to 94.1% of the world olive oil production, i.e. 2.63 million tonnes, if the countries of the southern side of the Mediterranean basin are included (Morocco, Algeria, Tunisia, Lybia, Egypt, Jordan, Israel, Lebanon, Syria and Turkey). Olive trees cultivation and olive oil processing, the latter step often taking place in locations distant from the olive production sites, generate every year in the northern side of the Mediterranean basin 6.01 million m^3 of liquid waste (**Figure 1**) and 8.06 million tonnes of solid waste (**Figure 2**) as an average. This amount rises up to 30 million m^3 of olive mill wastewater and 20 million tonnes of solid waste, called wet husks [6] if the southern side of the Mediterranean basin is

Figure 1. Ponds for the confinement of the wastewater (left = before filling, right = after filling) from the extraction of olive oil at the facility of Coop. Sor Ángela de la Cruz, Estepa (Sevilla, Spain). Due to the environmental toxicity of the liquid phase, the large amounts produced from intensive olive tree cultivation in that area need to be stored separately before further processing, e.g. for biogas production (Source: Marco Nuti).

Figure 2. Olive pomace (wet husks) from the two-phase decanter centrifugation olive oil extraction. The wet husks have an initial relative humidity of 65–70% (Source: Marco Nuti).

included. The farming system in the Mediterranean area consists of a majority of small farms (e.g. in Italy 60% of olive farms are <2 ha, and less than 10% are >10 ha), and olive trees are generally grown along with other cash and food crops, infrequently with livestock. Overall farm economics is dependent upon a wide variety of factors, including productivity, quality, and costs of production and waste disposal. To achieve quality and sustainability standards, the costs are on the premises of the farmers, while premiums are limited to larger agro-industrial installations transforming the waste into energy, i.e. electricity (**Figure 3**). The options for the valorization of the olive oil extraction solid residues, not focusing on waste management in rural areas, have been recently reviewed [8, 9].

Figure 3. Power generation plant (1 Mw) fed with wet olive husks and wastewaters from the olive oil extraction. Biogas is generated in the digestor (right) and stored (left), then converted into electricity. Installation Friel Ionica Srl - BTM Srl spinoff of the University of Pisa, located in Manduria (Taranto, Italy) (Source: Marco Nuti).

There are many similarities between the chemical composition and end-of-use of the solid residues of the two production chains, besides their vast amounts available yearly. Both solid residues (i.e. coffee husks, defective coffee beans and spent coffee grounds on one side, and olive wet husks, olive stones on the other side) are problematic in terms of disposal, having similar environmental impacts, decontamination needs, similar possible re-uses and similar chemical nature of the components, i.e. a high content in ligno-cellulosic materials, low content of fat and protein, presence of (poly)phenols recalcitrant to degradation. The main end-of-uses in the poorer and less accessible rural areas are the production of heat and recycling into agriculture after modest composting. Production of heat for household heaters, electricity, recycling into agriculture after modest composting and minor uses for the production of commodities with high added-value (e.g. cosmetics, mushrooms, fodder) characterize the end-of-use of the solid residues in areas with more intensive agriculture.

In this chapter, the two types of solid wastes are critically reviewed separately and assessed for their valorization in the rural areas for production of coffee and olive oil, respectively.

2. The management of residues of coffee production and transformation in rural areas

Over 90% of the coffee production (*Coffea arabica, Coffea canephora* and the Ethiopia's natural *C. arabica* cultivar Harrar) takes place in developing countries. In these countries, economy depends to a large extent on agriculture, coffee being one of the most important crops. In fact, countries like Vietnam, India, Kenya, Nicaragua, Ecuador and Mexico encouraged the cultivation of coffee in rural areas as a national economy strategy. About 70% of the world coffee production is cultivated in rural areas on small farms less than 6 ha [10]. The implemented policies, subsidies and incentive programs which promoted the conversion for land to intensive, technified mono-crop coffee cultivation resulted in a catastrophic impact to the environment. Coffee farms are located indeed in some of the biologically most diverse, and most threatened, environments in the world [11].

The isolated rural areas where the best coffee is grown are extreme poverty zones (**Figure 4**). They have limited or no basic services (**Figure 5**), access roads are in poor condition and farmers have little or any basic education. For coffee growers in many countries, coffee provides their sole source of cash income and it is a family activity. Farmers have relatively weak trade positions and, in spite of it, they have to hold the same high productions standards as the large-scale producers who have additional resources to invest and access technological tools. To join the forces and improve the marketing of coffee, marginalized farmers created small communities called 'cooperatives' or 'beneficios' that are collection centres where they gather the crops and process the fruit until obtaining the coffee beans. The process of separation of the commercial product (beans) from coffee cherries generates enormous volumes of waste material in the form of pulp, residual water and parchment. Almost all these waste are disposed in the natural environment, causing bad odours, bad aspect, pathogenic insects' attraction, and pollution of soils and water bodies. In fact, they represent the major source of river pollution in Ethiopia and northern Latin America [12, 13]. The appropriate use of coffee by-products would help circumventing these problems, and valorization of the residues would represent a value addition from the point of view of environment protection [14]. The main chemical traits of the coffee processing residues at farm level, relevant for their valorization,

Figure 4. The small 'coffee farms' in the Amazonas very often look like an orchard exhibiting great plant biodiversity (Source: Marco Nuti).

Figure 5. Coffee beans are air-dried by small farmers in the western Amazonian forest at 1200 m of altitude (facility of the 'Asociación de cultivadores y comercializadores de café organico Bosque Nublado Río Golondrinas ', Parroquia Jijon y Caamano, Carchi, Ecuador) (Source: Cristina Echeverria).

are summarized in **Table 1**. The degradation process, occurring in natural conditions, is extremely slow and incomplete despite a favourable C/N mass ratio of 40, due to the recalcitrance of (poly-)phenols and complex glucides, thus giving rise to toxicity problems.

Various attempts have been made to circumvent these environmental problems and minimize the toxicity levels, e.g. improving production systems, reducing the volume of wastewater or recycling wastes to obtain value-added compounds such as enzymes and caffeine [1]. Out of the alternatives, only a few have been implemented in rural areas because of the costs and unavailability of the technology in these small communities. The valorization of coffee residues in agriculture, as animal feed and for energy production is still the most attractive application to solve in part the problems of people in these areas.

During the years 2000–2004, many farmers started a transition to organic coffee production, encouraged by the growth of certified coffee markets and development projects. This transition involved the use of new forms of fertilization [11, 15]. From this point of view, large quantities of coffee pulp are available for those organic farmers. Approximately three-quarters of the nutrients extracted to obtain the coffee beans are found in the pulp. Co-composting the solid waste with animal manure is the most used alternative to minimize the environmental impact of the residue. However, in most cases the compost is not obtained under controlled conditions to ensure its stabilization and sanitization. Pulp is typically left to degrade in piles without any treatment, thus resulting in an organic material incorrectly named 'compost'. Introducing a non-stabilized organic matter into the soil has caused a negative effect on crops

http://dx.doi.org/10.5772/intechopen.69427

Component	Spent coffee grounds	Coffee pulp	Coffee husk
	g × 100 g^{-1} d.w.	g × 100 g^{-1} d.w.	g × 100 g^{-1} d.w.
Proteins	6.7–13.6	10.1	5.2–11
Total lignin	33.6	–	–
Cellulose	8.6–13.8	–	16.0–43.0
Carbohydrates	–	63.2	35.0–85.0
Reduced sugars	–	12.4	0.71
Ash	0.43–1.6	8.3	0.7–6.2
Fat	6.3–28.3	–	0.3–3.0
Tannins	1–9	1.80–8.56	
Caffeín	1–2	1.3	1.0–1.3
Organic carbon	–	–	50.8
N	–	–	1.27
Polyphenols	–	–	1.22

Data collated from different authors ([65–67] and references cited therein), showing a fairly wide range of values, possibly due to the diversity of tested materials, i.e. Arabica or Robusta variety, not specified by Aa. The ligno-cellulose component and tannins are the most recalcitrant components to degradation.

Table 1. Main chemical components of the solid coffee residues (spent coffee grounds, coffee pulp, and coffee husk) relevant for their valorization at small farm level.

production [16]. Vermicomposting is an alternative widely used in Colombia and Ethiopia showing better results in terms of organic fertilization [17].

Small projects to obtain bioethanol and biogas have been also implemented in some key areas of Africa and Central America [18–20]. The production of energy from coffee wastes partially meets the energy needs in rural areas. Furthermore, the technology needs further improvement in order to be applied in the most remote or marginalized places.

The use of pulp as an animal feed has had limited use due to the high content of caffeine which affects ruminants. In Ethiopia, it has been used as a nutrient supplement to sheep feeding [21, 22].

In conclusion, there is still a lot of pollution because of the lack of knowledge and government policies. There is a vital need to counterpart this production with an appropriate utilization of coffee by-products. The valorization of the liquid and solid residues should be regarded as a value addition from an environmental point of view. However, it is essential that coffee production and processing take into account environmental needs to ensure sustainability, reasonable living standards for the populations involved with coffee, and ensure the maintenance of quality. Such an effort is one of the objectives of the International Coffee Agreement 2007, i.e. to encourage members to develop a sustainable coffee sector in economic, social and environmental terms [23].

3. The management of residues of olive oil extraction process in rural areas

The olive tree cultivation in the Mediterranean countries of EU (ca. 400 cultivars in Italy, 150 in Spain, 40 in Greece) takes place in 4.8 million ha, and the olive oil extraction process is carried out in about 12,000 olive-mills, most of which are small and medium enterprises (SMEs), involving 800,000 jobs [6, 24]. Olive oil is obtained essentially *via* traditional pressing (TP), two-phase decanter process (2-PDP) and three-phase decanter process (3-PDP) [25], each of them generating different amounts of wastes (**Table 2**). There is an uneven distribution of the type of extraction process in southern Europe: in Spain 99% of the adopted technology is 2-PDP, while in Italy 55% is 3-PDP, 15% is 2-PDP and 15% is TP. In Greece, 82% of the olive mills have adopted 3-PDP and 18% 2-PDP. However, there is an increasing interest in these last two countries for the two-phase system, possibly coupled with the de-pitting of the olive wet husks (synonyms: crude cake or pomace). In this way the stones can be separated and used for heating purposes, particularly at small community level and for household heaters.

As far as the residues from the olive trees cultivation are concerned, in the rural areas of southern Europe, two main components of biomass burning are the incineration of wood as household fuel, and the combustion of crop residues (i.e. prunings and leaves of olive trees) in open fields. At the same time, as the population continues to rise in the African side of the Mediterranean basin, the contribution from these two types of biomass burning tends to

Extraction process	Input	Amount of input	Output
Traditional pressing	Olives	1 tonne	Oil
	Washing water	0.1–0.12 m^3	Solid waste (*ca*..25% water + 6% oil)
	Energy	40–63 kWh	Waste water (*ca*. 88% water)
3-Phase decanter (3-PDP)	Olives	1 tonne	Oil
	Washing water	0.1–0.12 m^3	Solid waste (*ca*. 50% water + 4% oil)
	Fresh water	0.5–1 m^3	
	Water to wash the impure oil	10 kg	Waste water (*ca*.94% water + 1% oil)
	Energy	40–63 kWh	
2-Phase decanter (2-PDP)	Olives	1 tonne	Oil
	Washing water	0.1–0.12 m^3	Solid waste (*ca*. 60% water + 3% oil)
	Energy	<90–117 kWh	

Table 2. Solid and liquid waste generated using different olive extraction technology. The solid waste contains olive pits [25].

increase. In the northern side of the Mediterranean basin burning olive branches and leaves in open fields is allowed (**Figure 6**) under an increasingly stringent legislation as they can contribute to greenhouse gas emissions (GHG). These burnings can be avoided once the agro-residues are employed for sustainable, cost-effective and environment-friendly options such as composting and subsequent ploughing of the compost. A quantitative description of the spatial distribution of biofuel and open-field burning has been attempted to assess the impact of this burning on the budgets of trace gases [26], but no real attempts to discourage burnings are currently made on the basis of continuous educational programs for small olive farmers. Therefore, open-field burning of olive crop residues is still the most traditionally adopted end-of-use in rural areas. From a purely agronomic standpoint, the delivery of ashes in cropped agriculture would be meaningful for fertilization purposes only when the content of soil organic matter and organic nitrogen at plough depth is high, which seldom occurs in both northern and southern side of the Mediterranean basin [27–29].

The use of in-house pressing (i.e. with large stone wheels or stone cones) of the olives in small farms has almost disappeared, and this type of extraction is often confined to demonstration farms for teaching or museum purposes. The traditional pressing is also decreasing in favour of the decanter-centrifugation systems, either two- (olive oil and wet husks) or three-phases (olive oil, husks and wastewater). The small olive mills in rural areas work preferably with 3-PDP, but the availability on the market of 2-PDP decanter extractors with reduced energy consumption is gradually offering new opportunities. On the contrary, in agriculture-intensive areas, the preferred technology is 2-PDP with large working capacity (**Figure 7**) or the last-generation decanter extractors combining the modern extraction technology without the addition of water with batch processing, thanks to the bowl discharging device (**Figure 8**). Using the latter system, the waste is represented by a dehydrated husk similar to the one coming from the three-phase decanter, along with the pulp from the husk, the so-called 'pâté', i.e. wet husks without any trace of kernel directly inside the bowl. This pâté can be used for various purposes, including agronomic use, animal feeding, or can be mixed with other biomass for biogas production.

The biochemical and physical-chemical traits of the solid waste 'wet husks' are reported in **Table 3**, where for convenience they are compared with the traits of the wet husks after micro-

Figure 6. Burning olive leaves and prunings in the open field in controlled (left) and uncontrolled (right) conditions. In the olive oil producing EU countries burning is submitted to stringent legislation (Source: https://i2.wp.com/www.quiantella.it/wp-content/uploads/2016/06/a.jpg?fit=533%2C348&resize=350%2C200).

Figure 7. Two-phase decanters in series in an olive mill for the extraction of olive oil (Picualia, Proyecto de traslado y perfeccionamento de almazaras por fusiòn de Cooperativas 'Agricola de Bailén virgen de Zocueca', Jaen, Spain) (Source: Marco Nuti).

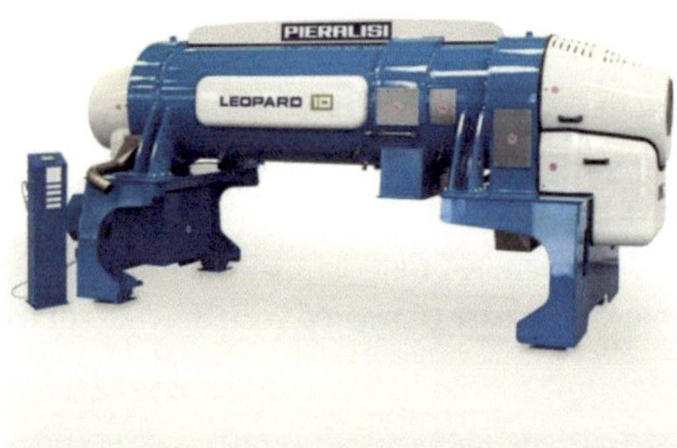

Figure 8. Last generation decanter extractors of olive oil, which combines the two-phase extraction technology without the addition of water with batch processing, thanks to the bowl discharging device. The installed power ranges from 7.5 Kw (more adapted to small quantities of olives to be extracted, i.e. process capacity of 0.5 tonnes/hour) to 45 Kw (for very large quantities of olives to be extracted, i.e. process capacity of 9 tonnes/hour) (Source: courtesy of Pieralisi SpA, Ancona, Italy).

bially enhanced composting in mechanically turned static piles. The re-uses at small farm level include (i) further oil extraction as the wet husk still contain 2–4% oil, (ii) delivery by soil treatment into the legally allowed soil acreage, (iii) sale to larger companies, (iv) re-use as fodder for animals, which requires pre-treatment with appropriate enzymes and process, (v) delivery into soil after composting, as a green (i.e. only plant materials) or mixed (i.e. plant

Parameter	Time (days)	
	0	60
pH	6.4aA	8.1bB
EC (dS m^{-1})	1.28aA	0.79bA
TN (UoW)	2.05aB	1.88bB
TOC (UoW)	46.8aA	31.9bA
C/N	22.8aB	16.9bB
HA (UoW)	15.6aA	41.6bB
FA (UoW)	14.3aA	19.5bA
HA/FA	1.10aA	2.14bB
HI	1.1aA	0.1bB
HD	48aA	94.5bB
HR	6.4aB	19.2bB
Ash (%)	8.3aA	11.6bB
Fats (UoW)	21.5aA	2.9bB
Lignin (UoW)	47.7	27.8
Phenols (UoW)	14.0bA	1.6 bB
Respiration (CO_2–C)mg g-1)	1087.8aB	678.3bA

UoW = units of weight. The microbially enhanced composting (static piles with mechanical turning) was carried out with the use of selected microbial starters [60]. Different letters indicate statistically significant differences ($P < 0.05$). Low case letters, comparison between sampling times within each treatment. Capital letters,comparison between treatments within each sampling time.

Table 3. Chemical-physical properties of the initial olive wet husks ($t = 0$) and of the stabilized compost after 60 days of composting ($t = 60$) in controlled conditions.

material plus manure) fertilizer. None of the first three options is considered really profitable by small farmers, because pomace oil has a low price with very marginal profit, direct delivery into soil requires a lot of bureaucracy and intoxicates the soil and selling the husks as a fuel is poorly profitable and needs transportation of a material with 70% humidity. If air-dried before transportation, the volatile phenolics can cause air pollution. The fourth option would be feasible, provided that the technology is available to the farmer. The last option is probably the most feasible in-house, provided the farm has a tractor for the mechanical turning of the piles.

4. Agronomic and legislative aspects

Agricultural utilization of residues as organic amendments and fertilizers has been shown to be a sound alternative for both residue recycling and soil fertility improvement [16]. The latter goal can be achieved in the frame of a modern agronomic management such as the conservative and regenerative agricultural practices. Regenerative agriculture stands on the three pillars of the conservative agriculture (use of crop rotation, reduction in tillage, retention of adequate levels of crop residues and soil surface cover) plus the maintenance of soil carbon sink. All these management practices can lead to a significant increase of carbon content in soil [30]. In turn, the increase of organic matter and humic fractions in soil determine the increase of soil richness and diversity of microbiota [31]. Therefore the utilization of residues of both coffee and olive cultivation, along with the utilization of the residues of the first step of processing (i.e. those feasible at small farm level in rural areas) cannot be merely identified with their disposal. On the contrary, the utilization of these residues is advisable, particularly because advantages to crops and soil are expected, either in the short- or in medium-term.

More specifically, the good agronomic practices (GAP) adopted for coffee cultivation by both top and low-producing countries, e.g. [32–34], define the criteria leading to a product conforming quality and safety criteria in regimes of both conventional and organic agriculture, and include the use of organic fertilizers and their quality. Though the aims are the same, the rules and recommendations can vary, reflecting the different pedo-climatic characteristics of the cultivation area. The density and productivity of coffee plants per hectare for small holders coffee farms can range from 1332 plants of Arabica, with a very low productivity of about 400 kg in the traditional organic coffee orchards of the Galapagos Islands (Ecuador) [35] to 1100 plants producing up to 3.5 tonnes per ha in Vietnam with Robusta variety grown with high farm inputs [36], or high yields in intensified monocultures with a density of 10,000 plants [37].

Maximizing the small coffee farms seems to be linked nowadays more to the quality of the beans rather than to the yields per ha. In this sense, enhancing the bean quality by minimizing or avoiding chemical inputs and maximizing the re-use of correctly composted residues could help in achieving the task. On the contrary, in intensive coffee cropping systems where the predominant criterion is the harvesting cost, the trend is to have much higher plantation density since it costs almost as much to harvest a low-yield as a high-yield field. But in this case, additional costs could emerge for shading management systems (arborization), for more irrigation inputs, and more plant protection products usage. For the legislative framework of organic fertilizers, biostimulants and microbial-based amendements in coffee-producing countries, most of them have installed recent rules for the safe use, production or import. As an example see the rules in Brazil [38], Vietnam [39] and Colombia [40] among the top producers, and Ecuador [41] among the smaller coffee-producing countries.

The olive tree is considered as one of the cultivated trees with the lowest demand for soil nutrients. This is the main reason why the tree can survive and be productive even in poor, rocky areas with soils mostly derived from hard limestone, e.g. in Greece, Italy and Spain, or in sandy soils in the southern side of the Mediterranean basin, e.g. Tunisia and Morocco. A significant portion of the olive groves can be found, in the small farms of the EU countries, on steep hill and mountain slopes which have been terraced with stone walls to hold the soil. For the olive chain residues, the amount of residues at farm level will be strongly dependent on the density of olive tree plantations. In the actual agronomic management of olive groves in the Mediterranean basin, the density of olive trees plantations ranges from 10 to 15 trees per ha of Tunisian or 40–50 trees per ha of Puglia's (Italy) small farms (due to low water availability) and soil often maintained without cover, to more than 1500 trees per ha in the intensive new cultivation areas of Spain and Italy. The inter-row space, in the intensive cultivation areas, is often left without cover in Andalusia, while many small farmers in Italy have continued to adopt the traditional intercropping of olive groves with vineyards and arable crops (**Figure 9**). In northern and central Greece, farmers have historically combined olive production with arable crops in the same plot. This practice is reputed to be appropriate to ensure a steady economic return year-after-year, irrespective of the weather conditions. The positive contribution of agroforestry mixed with olive groves include continued olive production along with benefits in terms of animal health, appropriate control of manure usage and the creation of wildlife habitats. In a recently started project in the province of Chalkidiki (Crete, Greece), the olive production that takes into consideration both biodiversity maintenance and wildlife habitats showed high performances, whereas the main negative effects included extra costs of management, administrative overburden, the complexity of the planning and field work and aspects related to mechanization [42]. Small farmers in Greece and Italy have identified that intercropping is probably only appropriate where the principal product is represented by the olives for olive oil production, rather than the edible olives which require a relevant use of pesticides. The presence of some understory species in the cropped area is thought to enhance both quality and flavour of the olive oil.

For the residues from olive cultivation and olive oil extraction, in the Mediterranean basin there is an increasing trend to frame the soil application with unprocessed residues into a more stringent legislation. At EU level the matter is regulated by the Waste Framework Directive 2008/98/EC [43], the Directive on Industrial Emission of 2010 [44] implemented by the European Commission in 2012 [45]. For landfill disposal during the whole life-cycle of the landfill, the relevant rules to prevent or reduce the pollution of surface water, groundwater, soil and air, and the resulting risks to human health, are provided by the Landfill Directive 99/31/EC [46]. The EU legislation on this issue has been critically reviewed [47]. At national level, in Spain the disposal of olive chain wastes is regulated [48]. In Italy, the disposal of olive wastes is regulated by the national Law n. 574 of 1996 [49]. In Greece, the disposal of olive wastes is regulated by the national Laws n. 1650/86 and 3010/2002. The present legislative status in Greece does not allow the application of untreated olive mill wastes to soil surface [47].

Figure 9. Extensive olive tree (var. Picual) cultivation (left) along the 'Carretera de los olivares' between Jaen and Sevilla (Spain). In the province of Jaen there are over 40 million olive trees. The olive cultivars (cv.) mostly grown in Andalucia are Hojablanca, Picual, Lechin, Picudo, Verdial, Cornicabra, Empeltre, Arbequina. In Tuscany (Italy) the small farms often grow olive trees (cv. Leccino, Moraiolo, Frantoio, Pendolino, Leccio del corno, Maurino) intercropped with vines (right) (Source: Marco Nuti).

	Process and essential components	Minimum content/useful substances	Obligatory to be declared	Notes
Green composted amendment	Product obtained through a transformation and stabilization process, in controlled conditions, of organic residues. These can be prunings, olive husks, crop residues, other residues of plant origin.	Max humidity: 50%	Humidity	The content of other forms of N, total P and total K can be declared.
		pH 6–8.5	pH	Plastics, glass and metals cannot be higher than 2%
		Minimum organic carbon: 20%	Organic C	Stony inerts (diameter ≥ 5 mm) cannot be higher than 5%.
		Humic and fulvic carbon: min 2.5%	Humic and fulvic C	*Salmonella*: absent in 25 g of the sample w.w. (where $n = 5$, $c = 0$, $m = 0$, $M = 0$)
		Organic N ≥ 80% of total N	Organic N	*Escherichia coli* lower than 1.000 cfu (where $n = 5$, $c = 1$, $m = 1000$ cfu/g, $M = 5000$ cfu/g)
		Max C/N: 50	C/N	Germination index (diluted 30%) ≥ 60%.
			Salinity	Algae and aquatic plants are allowed, such as Posidonia left on the shores, after separation from sand of the organic fraction. Their content must be lower than 20% of the initial mix.
			Na content	Thallium must be lower than 2 mg kg^{-1} (only in amendments containing algae).
Mixed composted amendment	Product obtained through a transformation and stabilization process, in controlled conditions, of organic residues. These can be by the organic fraction of USR from differentiated recycling of animal waste including liquid waste, residues of untreated wood processing and of the untreated textile industry, organic residues from effluents and muds, and all residues allowed for green composts.	Max humidity: 50%	Humidity	The muds (defined according the Legislative Decree 27 January 1992 n.99, cannot represent more than 35% (w/w) of the initial mix. The content of other forms of N, total P and total K can be declared. Plastics, glass and metals cannot be higher than 2%
		pH 6–8.5	pH	Stony inerts (diameter ≥ 5 mm) cannot be higher than 5%.
		Minimum organic carbon: 20%	Organic C	*Salmonella*: absent in 25 g of the sample w.w. (where $n = 5$, $c = 0$, $m = 0$, $M = 0$)
		Humic and fulvic carbon: min 7%	Humic and fulvic C	*E. coli* lower than 1.000 cfu (where $n = 5$, $c = 1$, $m = 1000$ cfu/g, $M = 5000$ cfu/g).
		Organic N ≥ 80% of total N	Organic N	.Germination index (diluted 30%) ≥ 60%.
		Max C/N: 25	C/N	Algae and aquatic plants are allowed, such as Posidonia left on the shores, after separation from sand of the organic fraction. Their content must be lower than 20% of the initial mix.
			Salinity	Thallium must be lower than 2 mg kg^{-1} (only in amendments containing algae).

All requirements are expressed in dry weight. The category 'Amendments' includes also manure, artificial manure, green non-composted amendment, composted turf, acid turf, neutral turf, humified turf, leonardite, vermicompost from manure, lignite. Cultivation substrates are in Annex 4, and the products with specific action on plants (e.g. mycorrhizal inoculants) are in Annex 6 of the same Legislative Decree [52].

Table 4. Specifications and requirements of the Italian Legislative Decree n. 75 of 2010 (Annex 2) for green composted amendments and mixed composted amendments.

Essentially, for the three major olive oil-producing countries the disposal is prohibited or strongly restricted in quantity, land area and timing. Unfortunately, the small farms in marginalized rural areas sometimes tend to overcome the restrictions, mainly because of the small quantities produced, the transportation costs and road difficulties. Also the production and quality of amendments, including those derived from a composting process of the olive cultivation and olive oil extraction process, are regulated by national laws in the EU countries of the Mediterranean basin. In Spain, the matter is regulated by the Fertilizer Act n.7540 [50] and n.13094 [51]. In Italy, the matter is regulated by the Annex 2 to the Fertilizers Act n. 75 [52]. As an example, the requirements of the Italian law for green and mixed amendments are reported in **Table 4**. Emphasis has been given to the source of materials to be used and the transformation process, to the physical-chemical traits of the amendments with particular attention for the level of humification, and to the hygienization and safety aspects. All the different types of amendments (non-composted, green composted, mixed composted) must conform to the limits of heavy metals, namely (in mg per kg dry matter: Pb 140, Cd 1.5, Ni 100, Zn 500, Cu 230, Hg 1.5, Cr^{6+} 0.5). In Greece, the matter is regulated by the Fertilizer Act n. 30(I) of 2006, and n. P.I. 118 of 2006. At EU level, the legislation on fertilizers, i.e. the Regulation (EC) No. 2003/2003 of the European Parliament and of the Council of 13 October 2003 [53], which was cantered on chemical fertilizers only, is actually being repealed by a new legislation that includes the organic fertilizers, biofertilizers and amendments. The approval of the new Regulation is expected by the end of 2017. The agronomic benefits from the use of a correctly composted amendment include a positive effect on soil structure, an increase of phytostimulatory substances, and a direct effect on crop yield. The latter is obtained through an increase of nutrient availability. In addition, as secondary effect it has been often observed that these amendments act as biosimulants or bio-effectors providing an increased biocontrol activity of soil-borne phytopathogens and a substantial soil detoxification. These traits may lead to some difficulty in placing these borderline products into an appropriate legislative framework [54]. The agronomic advantages of delivering green compost from olive waste as a fertilizer for olive groves include the possibility to run organic agriculture, to maintain and increase the soil carbon stocks and to detoxify the cropped area.

5. Perspectives of management practices

Different approaches are clearly needed to upgrade the residues of coffee and olive tree cultivation, as well as the processing residues. The variety of approaches is a consequence of the diversity of (a) pedoclimatic areas that include altitude and latitude, soil texture and organic matter content, water regime and availability, (b) level of expertise of the small farmers, (c) social environment that includes training opportunities and availability to create associative forms among producers, (d) access to trade and communication networks, (e) easy access to community-level processing installations.

In the case of coffee, the valorization of residues (**Figure 10**) in agriculture, as animal feed and for energy production, apart from a few minor uses, is still the most attractive application to respond to the challenges of the rapidly evolving socio-economic and poverty problems

Figure 10. Solid residues of coffee first step processing (i.e. up to the production of beans) in a small coffee farm in the Andean region (Ecuador). If not properly stored or quickly bio-transformed, the residue can be easily re-colonized by spoilage and pathogenic microorganisms (Source: Cristina Echeverria).

of the farmers in these areas. A minor use of plant leaves in organic coffee farms could be the production of herbal teas, whereas for the extraction of functional products for human food supplement, probably only more centralized processing installations can provide the appropriate machinery and food grade safety standards. Another minor use could be the production of edible mushrooms from (co)composted solid waste, i.e. mucilages and spent coffee grounds. This re-use has long been studied by Cenicafé in Colombia and interesting results were obtained with shiitake [55] and Pleurotus [56] as simple technology among low-income communities in the urban areas of the city of Manizales. However, sanitization and detoxification of the substrate remain the major problems and further development of substrate pre-treatment would help to obtain a mushroom production meeting food-grade safety standards.

The coffee prunings, actually mostly left *in situ* as a mulching agent or as an amendment, may retain their phytotoxity and presence of plant pathogens. Therefore their valorization *in situ* implies that they are processed via composting (or co-composting in farms where cattle manure is available). (Co-)composting will then be made by mixing prunings, leaves, coffee husk (i.e. the skin, pulps and parchment generated by pre- and post-fermentation de-hulling) and eventually manure. The residue amounts are relevant: for 1 tonne of coffee beans produced, ca. 1 tonne of husks are generated (dry coffee processing), while where wet processing is adopted, there will also be a relevant amount of wastewater from washings. The latter could be used to feed biodigesters for biogas production at community level since the process requires installations and machinery of capacity larger than the ones of a single small farm. In this case, transportation difficulties and costs should be taken into account.

The transformation of coffee plantations residues at farm level, along with the residues up to the production of the coffee beans, for agricultural end-of-uses, requires a remarkable improvement of the process as it is actually adopted in most cases by small farmers. The composting process of cultivation residues (prunings, leaves) should lead to the production on-farm of a detoxified, sanitized green composted amendment, suitable for use in organic agriculture, containing living microbial consortia. This goal can be properly achieved if these residues are bio-transformed together with the coffee processing solid waste. The actual composting process, far from being a science-based technology, needs the use of indigenous microbial starter cultures capable to degrade the recalcitrant substrates, to detoxify the phenolic toxic substances, transforming the intermediate and still toxic chemical compounds into useful phytostimulatory substances. In those farms where manure (preferably cow-dung and horse manure) is available, the process of co-composting should lead to a detoxified, sanitized mixed composted amendment. The two processes are different and require different expertises. In both cases, the use of selected starter cultures, bio-compatible among them, allows to include those microbial cultures having relevant phytostimulatory activity to the plants and also bio-control activity towards the most commonly encountered soil-borne pathogens. The use of the mature, sanitized, humified compost obtained in this way as a fertilizer could substantially contribute to strengthen the natural plant defence traits and therefore minimize the density of the soil-borne disease.

The second alternative in small farms could be the re-use of the solid residue as animal feed [57]. In different field trials, pigs and cows fed with up to 15% ensiled coffee pulp and 5% of bagasse showed no negative effects on weight compared to those fed with commercial concentrates, and the pulp used as a fodder in milking cows was shown to replace up to 20% of commercial concentrates. The advantages are that coffee husk and pulp are rich in glucides and minerals. However the presence of (poly)tannic complexes and of caffeine decreases the palatability of husk by animals. Furthermore, the caffeine has stimulatory and diuretic effects and tannins diminish the protein availability and inhibit digestive enzymes. By consequence, the removal of these two anti-physiological components would require pretreatments consisting in repeated washings and the use of commercial inoculants to enhance the fermentation (i.e. silage) process. Therefore this alternative looks less feasible economically at single farm level, and would be probably feasible at more centralized facilities level.

The third alternative of valorization could be the energy production. The use of biogas would fit for heating purposes at single farm level. Some case studies on the coffee processing factories indicate that the exploitation of the residues for the production of electricity is feasible. Studies carried out in Tanzania suggest that from coffee residues it is possible to obtain high methane yields: 650 m^3 of methane per tonne of volatile solids for Robusta variety solid waste and 730 m^3 methane per tonne of Arabica variety solid waste [58]. However, this alternative is probably more easily accomplished at a centralized facility level due to the engineering and expertise needed, rather than at single small farm level.

It appears, in conclusion, that for the small coffee farms the valorization of solid wastes are in any case tightly linked to initiatives of socio-economic nature, i.e. organize formal training and 'hands-on courses' for farmers, improve the road system and accessibility of the single farms, and facilitate the formation of 'cooperatives' among farmers.

In the case of olive tree cultivation waste, i.e. prunings and leaves, when they are still in the field can be finely cut and used as mulch (**Figure 11**) or ploughed into the soil. Another valorization would be to transform these residues into a humified compost. Recently a

Figure 11. A common agronomic practice in small farms for disposal of the olive tree prunings and leaves in situ: they are first placed inter-rows, then finely cut and finally used as mulch or ploughed into the soil (facility 'Azienda La Cerreta', Castagneto Carducci, Livorno, Italy). This practice represents a step forward compared to the burning practice and is considered cost-effective for small farms (Source: Marco Nuti).

composting process of prunings and leaves enriched with phosphate rock has been described in Saudi Arabia [59]. This bio-transformation process has a duration of 8 months, presumably because of the recalcitrance of the lignocellulosic substrate to degradation. In the case of olive oil extraction solid waste (wet olive husks), the most feasible option for small farmers is the re-use in agriculture through composting. This process is a knowledge-based technology, requiring some basic training for farmers. The process has been described by Echeverria et al. [60] at industrial level and can be applied also at farm level [61, 62]. Essentially it is a solid fermentation process carried out with the help of loaders for periodical turning the piles, and is different from static piles composting with/without forced ventilation. The biochemical transformation can be enhanced through the use of starters, prepared with virgin husks enriched with selected microbial cultures. The latter approach, with respect to composting without the use of the starter, allows to achieve deeper humification (i.e. higher content of humic substances), faster deodorization (disappearance of bad smells), shorter maturation time and better detoxification of the starting material. The process duration is, on average, 60– 90 days during which the initial material undergoes profound changes of its mechanical (e.g. particle size, texture), physical-chemical (e.g. pH, humidity, phenol/lignocellulosic content, humification indexes), and biological traits (e.g. sanitization of all potential human pathogens, appearance of bioactive phytostimulatory substances, different microbiological profile). Microorganisms are the main drivers of the transformations occurring in the substrate, and their degradation activity leads to the production of carbon dioxide and minor amounts of other gases which evolve in the atmosphere, and to the production of heat (which, if let uncontrolled, can easily go to >70°C

http://dx.doi.org/10.5772/intechopen.69427

leading to pasteurization of the matrix). From a biochemical point of view, the composting of wet husks must be viewed as a respiratory process which needs oxygen (the appropriate porosity and oxygen presence in the matrix is ensured by turnings and the presence of prunings) and gives rise to carbon dioxide. One of the consequences of the degradation of the substrate components with concurrent carbon dioxide formation is the loss of weight of the substrate, as an average 30–40% expressed in dry matter. Due to heat formation and periodical turnings, the water evaporates and as an average the humidity content decreases from the initial 65–70% to ca. 40% of the compost after 90 days. Complex biochemical reaction does occur too, which involves polycondensation and polymerization reactions leading to the formation of the humic substances useful for the soil fertilization purposes. The initial fresh matrix is toxic to plants, but after composting turns into a plant growth stimulator, due to the presence of auxins and other substances synthesized during the composting process and to the concurrent degradation of phenolic plant growth-inhibitory substances, both processes being of microbial nature. The success of the composting process will ultimately depend on (a) the initial quality of the wet husks and starters and (b) the ability of the operator(s) to maintain the appropriate process conditions leading to the formation of mature compost in the time limits. The conditions will consist mainly in keeping under careful control the main process factors, i.e. oxygenation, heat and humidity. Appropriate oxygen presence and heat control will be achieved through periodical turning the piles when the temperature rises above 50°C (indicated by long-stem thermometers). The appropriate humidity will be ensured through the addition of wastewater, i.e. 60–70% (initial humidity, before composting) to obtain a compost with ca. 40 % (final humidity). The mature compost can be delivered to the farm soils as such for fertilization purposes [63]. In alternative, humidity can be further adjusted by air-drying to the desired level. Humidity below 25% will allow longer storage of the product until use. The use of such a fertilizer on-site is highly compatible with the principles of the regenerative agriculture, i.e. provides the opportunity to maintain and increase the carbon stocks in soil at farm level. This goal, if achieved until the threshold value of organic carbon reaches the minimum value of 3.5%, which allows to maintain the functional soil biodiversity [64]. If, on the contrary, the wet husks are not composted correctly, they will retain their bad odour and phytotoxicity, along with little or any humification of the initial material. In addition to the above described advantages using a correctly made compost as a fertilizer, the presence of microbial consortia, having phytostimulatory activity for the plants besides their fundamental role in the biotransformation of the initial matrix, would help substantially to strengthen the plant natural resources, minimize the attack by soil-borne phytopathogens, and by consequence would allow the use of more eco-friendly land management approaches.

Acknowledgements

The authors wish to thank Prof. Laura Ercoli for critical reading of the manuscript. This work is part of a Research Grant (Proyecto Café) from UTN, Ibarra, Ecuador to MC Echeverria, PhD.

Author details

Maria Cristina Echeverria[1], Elisa Pellegrino[2] and Marco Nuti[2]*

*Address all correspondence to: mn.marconuti@gmail.com

1 Universitad Tecnica del Norte, General José Maria Cordova, Ibarra, Ecuador

2 Institute of Life Sciences, Scuola Superiore Sant'Anna, Piazza Martiri della Libertà, Pisa, Italy

References

[1] Echeverria MC, Nuti M. Valorisation of the residues of coffee agro-industry: Perspectives and limitations. Open Waste Management Journal. 2017;**10**:3-15

[2] Benin S, Thurlow J, Diao X, Kebba A, Ofwono N. Agricultural Growth and Investment Options for Poverty Reduction in Uganda. Washington D.C: International Food Policy Research Institute (IFPRI); 2008. Discussion Paper 790

[3] Hughes SR, López-Núñez JC, Jones MA, Moser BR, Cox EJ, Lindquist M, et al. Sustainable conversion of coffee and other crop wastes to biofuels and bioproducts using coupled biochemical and thermochemical processes in a multi-stage biorefinery concept. Applied Microbiology and Biotechnology. 2014;**98**:8413-8431

[4] Oliveira LS, Franca AS. An overview of the potential uses for coffee husks. In: Coffee in Health and Disease Prevention, Preedy VR, editor. Academic Press, Elsevier, Amsterdam; 2015. pp. 283-291. Chapter 31. http://dx.doi.org/10.1016/B978-0-12-409517-5.00031-0

[5] Ciesielczuk T, Karwaczyńska U, Sporek M. The possibility of disposing of spent coffee ground with energy recycling. Journal of Ecological Engineering. 2015;**16**:133-138

[6] Demeke, S. Coffee pulp alone and in combination with urea and other feeds for sheep in Ethiopia. Small Ruminant Research. 1991;**5**:223-231

[7] Cossu A, Degl'Innocenti S, Agnolucci M, Cristani C, Bedini S, Nuti M. Assessment of the life cycle environmental impact of the olive oil extraction solid wastes in the European Union. The Open Waste Management Journal. 2013;**6**:6-14

[8] Kalderis D, Diamadopoulos E. Valorization of solid waste residues from olive oil mills: A review. Terrestrial and Aquatic Environmental Toxicology. 2010;**4**(Special Issue 1):7-20

[9] Arcas N, Arroyo López FN, Caballero J, D'Andria R, Fernández M, Fernández-Escobar R, et al, editors. Present and future of the Mediterranean Olive Sector. In: IAMZ-CIHEAM, Options Méditerranéennes Series A: Mediterranean Seminars, Vol. 106. 2013. pp. 1-197. ISBN: 2-85352-512-0

[10] Eakin, H, Winkels A, Sendzimir J. Nested vulnerability: Exploring cross-scale linkages and vulnerability teleconnections in Mexican and Vietnamese coffee systems. Environmental Science & Policy. 2009;**12**:398-412

[11] Valkila J. Fair Trade organic coffee production in Nicaragua Sustainable development or a poverty trap? Ecological Economics. 2009;**68**:3018-3025

[12] Beyene A, Kassahun Y, Addis T, Assefa F, Amsalu A, Legesse W, et al. The impact of traditional coffee processing on river water quality in Ethiopia and the urgency of adopting sound environmental practices. Environmental Monitoring and Assessment. 2012;**184**:7053-7063

[13] Rice RA. A place unbecoming: The coffee farm of northern Latin America. Geographical Review. 1999;**89**:554-579

[14] Murthy PS, Naidu MM. Sustainable management of coffee industry by-products and value addition – A review. Resources, Conservation and Recycling. 2012;**66**:45-58

[15] Barham BL, Callenes M, Gitter S, Lewis J, Weber J. Fair trade/organic coffee, rural livelihoods, and the 'Agrarian question': Southern Mexican coffee families in transition. World Development. 2010;**39**:134-145

[16] Mazzarino MJ, Laos F, Satti P, Moyano S. Agronomic and environmental aspects of utilization of organic residues in soils of the Andean-Patagonian Region. Soil Science and Plant Nutrition. 1998;**44**:105-113

[17] Raphael K, Velmourougane K. Chemical and microbiological changes during vermicomposting of coffee pulp using exotic (*Eudrilus eugeniae*) and native earthworm (*Perionyx ceylanesis*) species. Biodegradation. 2011;**22**:497-507

[18] Calzada JF, De Porres E, Yurrita A, Cabello A. Biogas production from coffee pulp juice: One-and two-phase systems. Agricultural Wastes. 1984;**9**:217-230

[19] Pandey AC, Soccol R, Nigam P, Brand D, Mohan R, Roussos S. Biotechnological potential of coffee pulp and coffee husk for bioprocesses. Biochemical Engineering Journal. 2000;**6**:53-162

[20] Navia, DP, Velasco MR de J, Hoyos CJL. Production and evaluation of ethanol from coffee processing by-products. Vitae. 2011;**18**:287-294

[21] Givens DI, Barber WP. *In vivo* evaluation of spent coffee grounds as a ruminant feed. Agricultural Wastes. 1986;**18**:69-72

[22] Dermeche S, Nadour M, Larroche C, Moulti-Matia F, Michaud P. Olive mill wastes: Biochemical characterizations and valorization strategies. Process Biochemistry. 2013;**48**:1532-1552

[23] ICO, International Coffee Organization. Available from: http://www.ico.org/sustaindev_e.asp (Accessed: 15 April 2017)

[24] Bršcic K, Poljuha D, Krapac M. Olive residues – Renewable source of energy. In: Management of Technology – Step to Sustainable Production; 10-12 June 2009; Sibenik. Croatia. Available from: https://bib.irb.hr/datoteka/398136.dDokumenti_KristinaDocumentsRadovi2009-ibenikBrscic_Poljuha_Krapac.pdf

[25] MORE, Market of Olive Residues for Energy. SWOT Analysis [Internet]. 2008. Available from: http://www.moreintelligentenergy.eu/

[26] Yevich R, Logan JA. An assessment of biofuel use and burning of agricultural waste in the developing world. Global Biogeochemical Cycles. 2003;**17**:1095-1116

[27] Nuti M. Suolo, patrimonio dell'Umanità: Quanto ne stiamo perdendo per erosione, inquinamento e cementificazione? In: Academy of Sciences, Literature and Arts of Modena [Internet]. 2014. Available from: http://agrariansciences.blogspot.it/2014/11/suolo-patrimonio-dellumanita-quanto-ne.html

[28] Brahim N, Ibrahim H, Hatira A. Tunisian soil organic carbon stock - spatial and vertical variation. Procedia Engineering. 2014:**69**:1549-1555

[29] Dridi I, Arfaoui A. Organic nitrogen distribution in seven Tunisian soil types under contrasting pedogenetic conditions. Environ Earth Science. 2017:**76**:205. DOI:10.1007/s12665-017-6525-9

[30] Govaerts B, Verhulst N, Castellanos-Navarrete A, Sayre KD, Dixon J, Dendooven L. Conservation agriculture and soil carbon sequestration: Between myth and farmer reality. Critical Reviews in Plant Science. 2009;**28**:97-122

[31] Gomez E, Ferreras L, Toresani S. Soil bacterial functional diversity as influenced by organic amendment application. Bioresource Technology. 2006;**97**:1484-1489

[32] Pohlan HAJ, Janssen MJJ. Growth and production of coffee. In: Verheve WH, editor. Soils, Plant Growth and Crop Production. Vol. III. Oxford, UK: Encyclopedia of Life Support Systems, EOLSS Publishers; 2010. Available from: http://www.eolss.net

[33] Kuit M, Van Thiet N, Jansen D. Manual for Arabica Cultivation. Cam Lo, Quang: Tan Lam Agricultural Product Joint Stock Company; 2004. pp. 1-213

[34] BAFS, Bureau of Agriculture and Fisheries Standards. Code of Good Agricultural Practices for Coffee. Philippine National Standard. PNS/BAFS 169:2015 – ICS 67.080.20. Philippines: BPI Compound Visayas Avenue; 2015. pp. 1-30

[35] USDA, United States Departmnt of Agiculture. Ecuador Galapagos Coffee Production and Commercialization. Gain Report EC2013002[Internet]. 2013. Available from: https://gain.fas.usda.gov/Recent%20GAIN%20Publications/Ecuador%20Galapagos%20Coffee%20Production%20and%20Commercialization%20%20_Quito_Ecuador_1-25-2013.pdf

[36] Marsh A. Diversification by smallholder farmers: Viet Nam Robusta Coffee. Rome: FAO; [Internet]. 2007. pp. 1-35. Available from: http://www.fao.org/docrep/016/ap301e/ap301e.pdf

[37] Perfecto Y, Rice RA, Greenberg R, van der Voort ME. Shade coffee: A disappearing refuge for biodiversity. BioScience. 1996;**46**:598-608

[38] MAPA, Ministerio de Estado da Agricultura, Pecuária e Abastecimento, Brasil. Instrução Normativa no. 6 de 10/03/2016, Publicado no DO em [Internet].2016. Available from: https://www.legisweb.com.br/legislacao/?id=317445

[39] MARD, Ministry of Agriculture and Rural Development. Vietnam. Circular No. 36/2010/TT-BNNPTNT of June 24, 2010: Promulgating the Regulation on fertilizer production, trading and use. Available at http://extwprlegs1.fao.org/docs/pdf/vie98318.pdf

[40] ICA, Istituto Agropecuario Colombiano. Colombia. Resolucion no. 000698 del 4 Feb 2011 por medio de la cual se estabelecen los requisitos per el registro de departamentos técnicos de ensayos de eficacia, produtores e importadores de bioensumos de uso agricola y se dictan otras disposiciones. Available at http://www.ica.gov.co/getattachment/225bd110-d1c4-47d7-9cf3-43745201e39a/2011R698.aspx

[41] MAGAP, Ministerio de Agricultura, Ganaderia, Acuacultura y Pesca, Ecuador. Normativa general para promover y regular la producción orgánicaecológica-biológica en el Ecuador, Acuerdo ministerial no 299, Registro Oficial no 34 del 11 de julio de 2013. Available at http://www.agrocalidad.gob.ec/wp-content/uploads/pdf/certificacion-organica/1.Normativa-e-instructivo-de-la-Normativa-General-para-Promover-y-Regular-la-Produccion-Organica-Ecologica-Biologica-en-Ecuador.pdf

[42] Mantzanas K, Papanastasis VP, Pantera A, Papadopoulos A. Research and Development Protocol for the Olive Agroforestry System in Kassandra. Chalkidiki, Greece [Internet]. 2015. pp. 1-7. Available from: http://www.agforward.eu/index.php/en/intercropping-of-olive-groves-in-greece.html

[43] Directive 2008/98/EC of the European Parliament and of the Council of 19 November 2008 on waste and repealing certain Directives. Journal of the European Union, 2008. Available at http://eur-lex.europa.eu/legal-content/EN/TXT/?uri=celex%3A32008L0098

[44] Directive 2010/75/EU of the European Parliament and of the Council of 24 November 2010 on industrialemissions (integrated pollution prevention and control). Journal of the European Union. 2010. Available at http://eur-lex.europa.eu/legal-content/DA/ALL/?uri=CELEX:32010L0075

[45] European Commission. The implementation of the Soil Thematic Strategy and ongoing activities. Report from the Commission to the European Parliament, the Council, the European Economic and Social Committee and the Committee of the Regions, Brussels. COM(2012), 46 final [Internet]. 2012. Available from: http://eur-lex.europa.eu/legal-content/EN/TXT/PDF/?uri=CELEX:52012DC0046&from=EN

[46] Council Directive 1999/31/EC on the landfill of waste. Published on the Official Journal of the European Communities [Internet]. 1999. Available from: http://eur-lex.europa.eu/legal-content/EN/TXT/PDF/?uri=CELEX:31999L0031&from=EN

[47] Inglezakis VJ, Moreno JL, Doula M. Olive oil waste management EU legislation: Current situation and policy recommendations. International Journal of Chemical and Environmental Engineering Systems. 2012;**3**:65-77

[48] Del Castillo Quesada E, Pérez Giráldez MJ, Ruiz JCR, et al. Evolución de los sistemas de eliminación de residuos líquidos en almazaras. Higiene y Sanidad Ambiental. 2011;**11**:780-785

[49] Legge 11 novembre 1996, n. 574 'Nuove norme in materia di utilizzazione agronomica delle acque di vegetazione e di scarichi dei frantoi oleari' pubblicata nella Gazzetta Ufficiale (Italia) n. 265 del 12 novembre 1996. Available at http://www.parlamento.it/parlam/leggi/96574l.htm

[50] Real Decreto 506/2013, de 28 de junio, sobre productos fertilizantes. Publicado por el Ministerio de la Presidencia en el Boletino Oficial del Estado (España), n.164 del 10 de Julio 2013, Sec.I, pp. 51119-51207. Available at https://www.boe.es/boe/dias/2013/07/13/pdfs/BOE-A-2013-7713.pdf

[51] Ministerio de Agricultura, Alimentación y Medio ambiente. Orden AAA/2564/2015, de 27 de noviembre, por la que se modifican los anexos I, II, III, IV y VI del Real Decreto 506/2013, de 28 de junio, sobre productos fertilizantes. Publicado en el Boletin Oficial del Estado (España) Núm. 289 Jueves 3 de diciembre de 2015 Sec. I. pp. 114186-114248. Available at https://www.boe.es/boe/dias/2015/12/03/pdfs/BOE-A-2015-13094.pdf

[52] Legislative Decree n.75/2010. Riordino e revisione della disciplina in materia di fertilizzanti. Suppl. Ord. Gazzetta Ufficiale (Italia) n 106/L, Serie Generale n. 121, 26 May 2010. Available at https://www.compost.it/attachments/443_d_lgs_75_2010%20normativa_fertilizzanti.pdf

[53] Regulation (EC) No 2003/2003 of the European Parliament and of the Council of 13 October 2003 relating to fertilisers. Journal of the European Union. 2003. Available from: http://eur-lex.europa.eu/LexUriServ/LexUriServ.do?uri=OJ:L:2003:304:0001:0194:en:PDF

[54] Nuti M, Giovannetti G. Borderline products between Bio-fertilizers/Bio-effectors and plant protectants: The role of microbial consortia. Journal of Agricultural Science and Technology. 2015;**5**:305-315

[55] Fan L, Soccol CR. Coffee residues. In: Mushrooms Growers Handbook 2, Part I Shiitake, Shiitake Bag Cultivation. 2005. pp. 92-95. Chapter 4. ISSN 1739-1377. Published by MushWorld Haeng-oon Bldg. 150-5 Pyungchang-dong, Jongno-gu, SEOUL 110-846, KOREA. Available from: http://www.goba.eu/wp-content/uploads/2015/06/Mushroom_Growers_Handbook_2_-_Shiitake_Cultivation.pdf

[56] Jaramillo CL. Mushroom growing project in Colombia. In: Mushrooms Growers Handbook 2, Part II Mushrooms for Better Life. 2005. pp. 234-243. Chapter 9. Published by MushWorld. Available from: www.alohamedicinals.com/book2/chapter-9-02.pdf

[57] Bouafou KGM, Konan BA, Zannou-Tschoko V, Kati-Coulibally S. Potential food waste and by-products of coffee in animal feed. Electronic Journal of Biology. 2011;**7**:74-80

[58] Kivaisi AK, Rubindamayugi MST. The potential of agro-industrial residues for production of biogas and electricity in Tanzania. Renewable Energy. 1996;**9**:917-921

[59] Ghoneim AM, Elbassir OI, Modahish AS, Mahjoub MO. Compost production from olive tree pruning wastes enriched with phosphate rock. Compost Science & Utilization. 2017;**25**:13-21

[60] Echeverria MC, Cardelli R, Bedini S, Colombini A, Incrocci L, et al. Microbially-enhanced composting of wet olive husks. Bioresource Technology. 2012;**104**:509-517

[61] Echeverria MC, Cardelli R, Bedini S, Agnolucci M, Cristani C, Saviozzi A, et al. Composting wet olive husks with a starter based on oildepleted husks enhances compost humification. Compost Science and Utilization. 2011;**19**:183-188

[62] Agnolucci M, Cristani C, Battini F, Palla M, Cardelli R, Saviozzi A, et al. Microbially-enhanced composting of olive mill solid waste (wet husk): Bacterial and fungal community dynamics at industrial pilot and farm level. Bioresource Technology. 2013;**134**:10-16

[63] Gómez-Muñoz B, Hatch DJ, Bol R, García-Ruiz R. The compost of olive mill pomace: From a waste to a resource – Environmental benefits of its application in olive oil groves: In Sustainable development – Authoritative and Leading Edge Content for Environmental Management. pp. 459-483. Chapter 20, Sime Curkovic (Ed.), InTech, DOI: 10.5772/48824. ISBN 978-953-51-0682-1, Published: August 1st, 2012. Available from: http://www.intechopen.com/books/sustainable-developmentauthoritative-and-leading-edge-content-for-environmental-management

[64] Lynch JM, Benedetti A, Insam H, Nuti MP, Smalla K, Torsvik V, et al. Microbial diversity in soil: Ecological theories, the contribution of molecular techniques and the impact of transgenic plants and transgenic microorganisms. Biology and Fertility of Soils. 2004;**40**:363-385

[65] Caetano NS, Silva VFM, Mata TM. Valorization of coffee grounds for biodiesel production. In: Chemical Engineering Transactions. 5th International Conference on Safety and Environment in the Process, Cozzani V, De Rademaeker E., guest editors.Vol. 26. 2012.AIDIC Servizi S.r.l., Milano, Italy. ISBN 978-88-95608-17-4; ISSN 1974-9791. DOI: 10.3303/CET1226045

[66] Dzung, NA, Dzung TT, Khahn VTP. Evaluation of coffee husk compost for improving soil fertility and sustainable coffee production in rural central highland of Vietnam. Resources and Environment. 2013;**3**:77-82

[67] Bondesson E. A nutritional analysis on the by-product coffee husk and its potential utilization in food production – A literature study. Bachelor Thesis in Food Science. Uppsala: Publikation/Sveriges lantbruksuniversitet, Institutionen för livsmedelsvetenskap; 2015. p. 415. Available from: http://stud.epsilon.slu.se